Do you nose?

Americans use more than 200 billion tissues per year.

Aromatherapy dates back to ancient Greece when children were tucked into bed with lavender beneath their pillows to induce sweet dreams.

Most portraits of George Washington rendered him in profile since his "Roman" nose was heralded as indicating physical strength and inner wisdom.

The prophet Mohammed (570–632) named his top three earthly loves as "women, children, and perfume."

Versailles was known as *"la cour parfumé"* ("the perfumed court") since Louis XV demanded a different fragrance for his chambers every day.

Sigmund Freud turned to cocaine for treatment of his sinuses, calling it a "magical drug" for its mental and physical effects.

Humans today can detect up to 10,000 different odors.

The Nose

"A witty, compassionate look at all things nasal."

—*The Oregonian*

The Nose

A Profile of Sex, Beauty, and Survival

Gabrielle Glaser

WASHINGTON SQUARE PRESS
New York London Toronto Sydney Singapore

The ideas, procedures and suggestions in this book are intended to supplement, not replace, the medical advice of trained professionals. All matters regarding your health require medical supervision. Consult your physician before adopting the medical suggestions in this book as well as about any condition that may require diagnosis or medical attention.

The authors and publisher disclaim any liability arising directly or indirectly from the use of this book.

Washington Square Press
1230 Avenue of the Americas
New York, NY 10020

ISBN: 0-671-03863-X
0-671-03864-8 (Pbk)

First Washington Square Press trade paperback edition September 2003

10 9 8 7 6 5 4 3 2 1

Manufactured in the United States of America

For information regarding special discounts for bulk purchases, please contact Simon & Schuster Special Sales at 1-800-456-6798 or business@simonandschuster.com

To Ilana, Moriah, and Dalia,
who have exquisite noses

contents

one

Memoir of a Nose

When I was growing up, I learned very early that I had—or was it that I developed?—an unusually acute sense of smell. This was often a blessing—I was able to sniff out hot wires before they caught fire, predict weather changes, and detect food on the verge of going bad. But my fifth sense also predisposed me to worry. Even before normal cues told me things had gone astray, my nose sensed danger, or fear. When I was a child, my father once severed a finger with a circular saw. I could smell his blood as soon as he rushed into the house, long before I saw the scarlet pools staining the wooden floor, or the tight panic of my mother's face. My sensitive nose could also make me sick: heavy perfume, diesel fuel, even the halitosis of someone nearby—had the ability to make me nauseous, or give me a migraine.

My parents had four children, and we lived on a beautiful but isolated farm overlooking Oregon's Cascade Mountains. In the countryside, smells mark seasons—of labor, of rest, even of grief—as neatly as the calendar organizes months. One wet April, the air was weighted with the sweet, cloying scent of hyacinths. They bloomed in a nearby pasture, unmindful that just steps away the man who had planted them was dying. As he lay motionless, my father, his only son, picked handfuls of the short stocky stems and brought them to the old man's railed bedside in jam jars. His ashen face sank into the pillow, and he labored to breathe. Silently my father pushed get well cards and pill bottles to one side of the nightstand, and gently set down the purple blossoms. My grandfather's nostrils fluttered to life like eyelids flying open after a startling dream. With his bruised yellow hand he reached to grasp my father's outstretched palm, and his head turned slowly toward the flowers. "Bring them closer, son," he rasped. I stood in the corner, too scared to go near.

That fall, we dropped some of those bulbs into the ground near his grave. Symbols of spring, they summon instead a blur of stale dark hospital air, chiseled granite, and my father's red-rimmed eyes.

It was my father, after all, who taught me to smell. He is an articulate man, but words are not his language. And so I became fluent in the same way all children learn to communicate, by mimicking. On mornings when he'd take us to school, he would pause a moment before opening the creaky door to his pickup, coffee mug in hand, and inhale the air around us as if he were breathing in life itself. "Smell that," he'd command,

and we would dutifully sniff, both trying to please and grasp the importance of what was unseen and unheard.

My father, a farmer like his father before him, had a devotion to the land so fierce you'd think God had given it to us and us alone, granting the deed at Mount Sinai. He fortified his ties to the farm with rituals, as if to convince him that what was ours would remain so forever.

Over time I learned just what he meant as he drew his great breaths, trying, fleetingly, to catch smells that the breeze would throw first one way and then the other. For odors told you when crops, from wheat to rye grass, were ripe; one whiff of morning air could convey that it was still too damp to run the combines, or too dry to fertilize.

On days when conditions were just right, growers ignited their harvested fields to kill pests and burn chaff. Naturally, those unattached to farming complained about the stink—and danger. Smoke from those fires could, and did, easily traverse a highway with a slight change in wind, obscuring drivers' vision and causing fatal pileups. Yet whenever the newspaper carried stories of peoples' concerns, my mother, angry at any hint of disrespect toward farmers, would throw it down in disgust. Even when great black plumes swirled into the bright August sun and bits of scorched straw floated onto our clothesline, she would be righteous in her indignation. "Smells like money to me," she'd mutter, to no one in particular.

For with my mother, too, scents had totemic value. But they didn't represent our livelihood, or tell us when to harvest crops. They symbolized something much more esoteric: her state of mind.

In the winter, I could tell upon waking when the creek separating our house from humanity was about to overflow. When it rose in the rainy months, no one could come in, and we could only get out—to school, to the grocery store—in a rusted old four-wheel drive. With the swollen water came a sense of dread. To smell the wet, murky air and the muddy, mossy creek was always to know that my mother's mood would soon deflate.

Food smells, though, were the biggest portent of all.. My mother, a product of the 1950s, was raised to be a lady, to wear nylons in airports and never raise her voice. She told us frequently we were her career; she didn't need another. Trained as a teacher, she was intellectually curious. But there was little time for reading if we were her job. And, of course, there was also the matter of helping to run a 4,000-acre farm. During harvest, she fed a hungry crew—a responsibility that wasn't so bad on good days. Slender neck bent over steaming pots, strong hands kneading bread dough, she would disappear the way a person gets swept away in an engrossing novel, unmindful that hours have passed.

But sometimes the pressure of keeping house, rearing four kids, and worrying about the weather (on farms, it's always an obsession) caught up with her. The kitchen, my mother's sacrosanct refuge, with its marble slab and gleaming stainless steel sink, could also be a prison, both hers and ours. Whether it was hormones, a tiff with my father, or one child too many with the flu, something would tip the scale between happy and not, and the only way out for her would be to bake: sour cream chocolate cakes, cinnamon rolls, Nestlé's Toll House cookies she stacked onto metal racks. On the way home from school, I'd get a sweet whiff of melting chocolate or an earthy waft of sourdough bread through an open window. My palms would sweat and my jaw would clench. For me, those warm, loving smells, which were intended to please us all, spelled peril.

Once, I opened the door to find my adolescent logic put to a very adult test. My brother, a childhood asthmatic, was alone and prostrate on the couch, struggling to breathe. Acrid clouds of sugar-turned-carbon billowed from the oven, and my mother sat slumped over the counter, fingers splayed in her dark hair (overindulging in chocolate had triggered a migraine). She seemed unaware of everything, except her very real need for Darvon. I still don't remember what I did first: thwart a fire, attend to my brother's breathing, or track down, by two-way radio, someone who could get my mother to the hospital for painkillers.

In the summertime and on rainy weekends—especially on rainy weekends—my mother would take my sisters and me to my grandmother's beauty shop, a squat white building near the county airport. We all had long blonde hair, but mine was the most "difficult." It hung down my back in thick straw-colored bumps. "Can't make up its mind to be curly or straight!" Nana exclaimed, as if it were a character flaw. So she tried to help it decide, and squeezed me in between her other appointments. She would iron it so that it hung in curtains around my face, the faint burn of dried ends settling, deadened, into my nostrils. Or she'd spiral it into pin curls with her fingers, dipping her pointy comb into a jar of green setting lotion that smelled like the fake evergreen of tree-shaped air fresheners.

Once, she insisted on giving us all permanents. She wrapped our necks with scratchy white paper and secured plastic capes around our shoulders, then led us, one by one, through the row of dangling pink plastic beads that separated one room from the next. They swayed, clicking gently, for as long as it took to get our hair washed. Afterward, my hair reeking of permanent solution, I sat in exile under a hot plastic dryer, air blasting from its tiny holes. Women talked and laughed so hard they slapped their thighs, but you could never hear what they were saying over the drone of the dryer. So I thumbed through forbidden old copies of *Cosmopolitan,* and watched the planes take off and land like so many rickety insects.

The air of that shop was a swift departure from the gentle smells on the farm. There was nothing natural inside—no fresh blackberries, no wood stoves crackling, no delicate roses in bud vases. Nana loved smells as far from the outdoors as you could get. She misted herself with White Shoulders each morning, and slathered peach-scented lotion on her reddened hands as her customers were "drying" in their pink Naugahyde recliners. On her break, she'd sit down, put her feet up, and pass around her pack of Virginia Slims. The women would flick cinders off their magazines into the tiny steel ashtrays embedded in the arms of their seats.

Smoke, nail polish, neutralizer, hair dye, bright blue barbicide. The smells were harsh and raw, but they were also freedom. If the farm felt like a bed-and-breakfast, Nana's shop was like a speakeasy. Women came to smoke and change their hair with permanents and peroxide. They'd come on payday (the fifteenth and thirtieth were always booked), after their shifts at the paper mill. Often, they had bad husbands, bad kids, or both, and my grandmother dispensed advice as freely as she fogged fresh curls with hairspray.

Some days, rank sulfur from the mill blew our way, overtaking even the chemicals inside the shop. But the worst was on muggy afternoons in summer, after a rain. A taxidermist worked in a converted garage down the road, and the stench of decomposing elks seemed to seep even through the walls of the shop. On such afternoons Nana lit incense she got from the Import Plaza and waved it around with the fervor of a preacher before the damned. But nothing was a match for the stink of rotting flesh, especially not cheap jasmine.

Smells, and the memories they inspire, have come back to me in inexplicable ways. On several occasions I've been taken aback by how powerful their effect can be. One hot, miserable morning in Manhattan nearly twenty years after I left my little hometown for good, I was on my way to give a speech to a group of bankers and psychologists. I hadn't prepared what I was going to say, and I began to feel very anxious as I plodded along the sticky streets. I passed a subway station and got a pungent whiff of urine mixed with roasting hot dogs and the sickening sweetness of sugar-roasted peanuts. A woman yammering into her cell phone walked by, leaving gusts of Eternity in her wake. I was really starting to panic when a Dominican woman opened the door to her crowded beauty salon. A waft of familiar odors—dye, acetone, and neutralizer, came my way; inexplicably, I felt better. I forgot about my speech, and wandered back, to those long afternoons in Nana's shop where lunch was a raspberry Pop-Tart and a can of Fresca, as my hair, tethered to curlers, dried.

One spring day when my older daughters were little, I had them both with me in the supermarket. As we rolled past the cheap floral display—it was decked out for Easter—my eldest daughter stopped to sniff some purple blooms that were nose-high to a three-year-old. "Pretty," she said, and thrust a plastic pot in my face. No sooner had I taken a whiff than tears sprang to my eyes. For more than twenty years I had avoided hyacinths as a painful souvenir of my grandfather's death.

And a few winters ago I was walking up the hill to my house in New York on a cold Saturday. The kitchen window was cracked open, and a warm baking smell was drifting onto my suburban street. It was far from the winding gravel driveway of my youth, but the smell was the same, and my jaws tensed immediately. Before I knew it, my gait had quickened to a jog despite the bags I was carrying. I burst through the door, expecting disaster, only to find my husband and daughters giggling as they stole bites of slice-and-bake cookie dough. Only then did I realize why I, who love to cook, never bothered to bake. I had always blamed my indifference on the tedium of having to measure precisely. But the smell of that warm chocolate suddenly made clear why I had always avoided making cookies and cakes.

While smells, for me, were paramount to navigation, my nose was also influential in other matters. As the product of several diverse gene pools, it set me apart. In my sparsely populated universe, most people were descendants of Scandinavian and German immigrants, with swimming pool–blue eyes and unassuming noses. My forebears, however, were slightly more exotic: my paternal great-grandparents, Polish Jews, had obscured their heritage so that their children might better fit in with their neighbors, and an American Indian grandmother on my mother's side passed herself off as a "gypsy." All this helped explain an obsession with the size of our noses, a vanity that seemed out of place amongst the abstemious Yankee and French Canadian ancestors we were purported to have. In any case, those disparate strands of DNA seemed to have a hand in shaping my nose into a remarkably unique appendage. By the time I was twelve, it was long enough so that, like a roll of film taken with a finger hovering over the camera lens, it was always in sight. But its shape didn't really vex me so much—I look a lot like my father, and I admired *his* nose. I did begin to take notice, however, once others let me know they had. In seventh grade I entered a teacherless classroom to see a small blond boy writing my name on the chalkboard. "Ask Gabrielle," he had scribbled. I stared at the writer's back, my heart fluttering as I waited to see what else he would say. When he turned away, the boy grinned at the class, his mouth glinting all silver braces. "She Nose It All," the words read. Students burst into laughter, and I slunk to my desk, trying hard not to let anyone think I cared.

Throughout my adolescence, I consoled myself with tales of famous oversized noses. Jimmy Durante was so taunted by schoolmates that he vowed never to make fun of anyone with crossed eyes or big ears. Judy Collins, as a child, hated her nose so much she slept with a clothespin pinched to the tip of it, like Jo in *Little Women.* I had never heard her sing, but I adored

Maria Callas after I saw a picture of her glorious profile in *Time* magazine. And my heart nearly leapt out of my throat when I read that Diana Vreeland said that people only liked small noses because they reminded them of "piglets and kittens." A strong face, she declared, had a nose with a real bone in it.

But celebrities were far away. I might have shrugged off indignities like the one in Mrs. Durfee's class if my nose had actually functioned well, but it didn't (and doesn't today). My genes also left me prone to all kinds of respiratory infections, especially sinusitis. My sick nose affected my breathing, my sleeping, and my time—cumulatively, I've easily spent a year in doctors' offices—and eventually deteriorated to the point of total dysfunction. I am one of millions of Americans who have had repeated sinus surgery (four). I've been hospitalized for serious sinus infections, and have spent weeks wandering around with lines stuck in my veins for intravenous antibiotics. I'm quite healthy otherwise (I've been checked for every possible frightening disease), but I shy away from people with the sniffles as if they had drug-resistant TB. For me, an ordinary cold can turn into an ordeal.

The most troubling aspect of my dysfunctional nose was the abrupt disappearance of my sense of smell. My sinuses were so scarred and swollen after one operation that for two years, no odor molecules could get where they needed to go. I was depressed and lost—"smell-blind," as the ancients called it; for me, the period seemed a particularly cruel curse. I couldn't smell my children, my husband, my favorite lilies, or the food I loved to cook. Words, somehow, failed to describe my loss.

I was lucky. My ability to smell eventually returned, but

only after several rounds of powerful steroids. One day the aroma of the garlic I was sautéing slowly, surely lifted into my nostrils. I was overjoyed—I could *smell* again—but my reaction was so primal that language didn't accompany the realization. What was it about the nose, this feature that doubled as a sense and an organ, that was so central to my life—and to most of humanity? As I thought about the nose's interplay in my life, I began to see its prominence everywhere. Consumers could buy incense, candles, and room sprays even while shopping for jeans. From Gogol to *Tristram Shandy,* the nose loomed large in literature. It was an obvious feature in pop culture and politics. Bill Clinton's proboscis was the butt—and focus—of thousands of caricatures. Jennifer Grey told TV viewers and magazine readers that her nose jobs had left her a stranger to herself.

Suddenly, science was investigating the nose as if it were a hidden body part that had only recently been discovered. New drugs administered through it could stop migraines and the flu in their tracks. Women whose noses picked up the sweat of other women began menstruating in sync. Neurologists looked to the nose's olfactory neurons—the brain's only contact with the outside world—for larger clues about our gray matter.

The more I learned, the more I began to realize that the understanding of the nose, unlike that of the heart or the lungs or the skin, was in its infancy. Its reign over our memories, its rule over beauty, and its role in our health, were only beginning to be decoded.

My quest to understand the nose led me on a three-year odyssey that went from perfumers on Madison Avenue to dusty medical archives, and from tony plastic surgeons' offices to research labs across the United States. As I retraced the steps of the researchers who have struggled to understand the workings of the nose, I came to see how little we truly know about this essential organ.

Some of what I ran across seemed banal, at least on the surface. For example, one-third of American adults picks their noses at least once an hour. Despite its inclination for third-grade yuks, nose-picking has grave implications. The touch of a contaminated finger to the eye or nostril is a primary culprit in spreading bacteria and viruses—ever more serious as germs become deadlier.

And some of what I found was shocking. As recently as the 1940s, doctors at Cornell Hospital in New York City treated their sick sinus patients with a variety of unusual approaches. In one they thrust electric prods up their subjects' noses to see how much "discomfort" they could tolerate. (The data later went into an oft-cited study on headaches.) One man, disabled by severe allergies, sought help from the doctors for his chronic sneezing. Rather than treat him with antihistamines, which were available at the time, they stuffed his nose with ragweed pollen, to "see" if he was indeed allergic. (He was; it took hours for his respiratory system to recover.)

My research also took me to the Mayo Clinic, where I met doctors who believe that the cause of chronic sinusitis—from which one in seven Americans suffers—may well be tiny, easily vanquished fungi. And on Manhattan's Upper East Side, where the rich and famous flock for nips, tucks, and facial overhauls, I met a plastic surgeon who was obsessed with the nose. He had grappled with his own large one for thirty-nine years before finally "fixing" it. His experience—and anguish—of traversing the role of doctor to patient gave me a special glimpse into the history, culture, and implications of changing our faces.

Researchers devoted to the study of olfaction, or smell, now know more about what happens when you take a whiff of coffee, or drive past a fish cannery, than ever before. New knowledge shows how important smell is—or at least was, evolutionarily speaking—to our very existence. Progress may have obscured the necessity of smell in surviving—now chirping phones, vibrating beepers, and red e-mail flags seem far more pressing than any odor—but eons ago, it was our species' lifeline. Science has now identified the initial path of synapses when we sniff, and how messages are transmitted to the olfactory bulb and olfactory cortex. But it is still pondering the larger question: exactly how tiny odor molecules prompt us to act, think, and remember.

In order to put it all in perspective, I had to turn to the distant past. I started with a gentleman whose training was both historic and cosmic: my rabbi, Lawrence Perlman. Rabbi Perlman is a tall, imposing man with elegant suits and a complex set of passions; Judaism, good food, good wine, and hockey (he played professionally). When I mentioned my research, he said the nose was threaded throughout Jewish texts, and invited me to come for a discussion. Our appointment was on a cold day in March, just as spring was beginning to flirt. Given our subject, I thought I'd bring something thematic along, so I sliced a couple of branches of daphne off the plant outside my window—its scent, a cross between lemon and cinnamon, is my favorite—and put them into a small vase.

When I got to the office, I set the glass near the window and sat down opposite the rabbi's giant antique desk. I could hardly keep up with his thoughts as he reached for first one and then another book marked with Post-its. We traversed texts from Genesis to Maimonides, and topics from respiration to odors, nasal shapes and nose rings. God created man by blowing into Adam's nostrils "the soul of life" (Genesis 2:7). The Hebrew word for breath, *neshima,* shares a root with *neshama,* soul. A Hebrew expression used to describe wrath, *haron-af,* evokes the nose: *haron* means rage, and *af* is the word for nose; together, they paint an image of someone so furious their nostrils are flaring. Only men with prominent noses could be chosen as priests. Even God, the rabbi said, had favorite smells.

As the weak afternoon light waned, the heat clicked on and the spicy scent of the daphne, perched above the radiator, infused the office. It seemed a fitting backdrop, as the rabbi began to sneeze. Between Kleenexes, he found a reference in the Talmud even to the meaning of sneezes.[1]

"The nose," said Rabbi Perlman, "is never just a nose."

And he is right. It moves through art, science, and popular culture from the sacred to the profane, and back again. From hieroglyphics to modern medical journals, the nose has been both an enduring mystery and an obsession, as fascinating to Pliny as it was to Picasso.

Part One

History of the Nose

The First Two Millennia

two

Centuries of Stench

In the beginning, the world stank. That, of course, is not surprising. The human animal is a powerfully industrious machine for producing odors—breathing, running, breeding, eating, defecating. Ancient man's first goal was to get in from the elements and learn how to use fire. Not much longer afterward—recorded history doesn't tell us exactly when—people turned to the problem of conquering smells, specifically their own. But odor was an enemy that, it turns out, was not easily vanquished. Unlike saber-toothed tigers and woolly mammoths, which could be stabbed with sharpened spears, the stench of man was a more complex problem.

In our modern, deodorized world, it's hard to imagine exactly how foul life could have been. Consider the ancient world, where almost everything people did created smells. Animals were roasted in entirety, skins, greasy entrails, and all, so cooking smells were a far cry from the cleaned-up, hairless barbecues we know today. Tents were made of bark, reeds, or wool, not some quick-drying microfiber. Just try soaking a woolen blanket with a garden hose and leave it outside for a month or two to get an idea of what would have been a mere minor assault to your olfactory system.

But smells did not only exist to menace. While the stench of life before sewers was a constant reminder of man's mundane needs, fragrance, on the other hand, was a powerful conduit to the esoteric. Smell could seduce lovers, cure the sick, and most importantly, link man to the divine.

Placating the Gods

The ancient Egyptians believed all pleasant odors derived from the tears and sweat of their many deities, and they reasoned that fragrance helped to underscore the importance of their prayers. And so the first perfumers used crushed seeds, roots, petals, wood, and fruit of the many aromatic plants found in the fertile Nile Valley, and mixed them with animal fat or oil. They were used as incense or unguents, sticky creams in an oily base. As early as the twelfth century BCE, formulas for these primitive perfumes list dozens of ingredients—lilies, dill, marjoram, and iris, all native to Egypt—as well as complex techniques for making them. Fragrance was so important that it was among the first items traded: by 3000 BCE, foreign roots, barks, and resins, which were all easily transported, appeared in the recipes as well. They included frankincense and myrrh, resins from small shrub-like trees in Somalia and Arabia; cinnamon bark from East Africa, juniper from Syria, camel grass, a rose-scented plant from Lebanon, and mastic, a fresh-smelling bush from Greece.

The name of the most popular Egyptian fragrance, *kyphi,* means "welcome to the gods." It had a variety of uses, from decongesting stuffy noses to coaxing sleep. Priests slathered themselves with it before visiting temples. There, they used it to "awaken" the gods by wafting it under the statues' noses.[1]

Everyone from noblemen to slaves bathed daily, and used aromatic oils and ointments liberally. Perhaps the fixation with fragrance was born of good reason: there was no soap. (The first soap, a crude mixture of ash and animal fats, was produced in ancient Babylon in 2500 BCE.) Though it is unlikely that any amount of fragrance could offset the body odor generated by the intense heat, the Egyptians made various attempts at the first deodorants. One method called for wadded up balls of pine resin that were then placed in the armpits.[2] Of all the remarkable techniques the Egyptians used to scent themselves, perhaps the most peculiar was what Egyptologists call "unguent cones." On special occasions, people wore an oily, waxy mass of semisolid perfume on top of their heads; they eventually melted, of course, and streamed down the face and neck in slick trails (apparently, they were thought to be cooling).

Scent was thought to carry the dead along on their journey to the afterlife, too. Bodies were embalmed with herbs and oils; even perfume trays were buried along with mummies for the trip to the next world. When Howard Carter finally cleared King Tutankhamen's tomb in the Valley of the Kings in

1922, he and fellow archeologists found a bowl of frankincense inside. They burned a sample, and found that even after 3,500 years underground, it gave off a "pleasant, aromatic odor."

Smells of the Book

In ancient Israel, riverbeds were so paltry that rainfall was stored and collected for irrigating crops, not people. Waterproof cisterns weren't developed until 1550 BCE; until then, people settled near springs or wells. Without water to drink or to bathe in, human smells took on powerful meaning. Bad breath was so common—and offensive—that by Talmudic times (third century BCE–700 CE) the rabbis listed remedies for it—chewing the bark and resin of the mastic tree.

Bodies themselves fared little better. Much of Israel is humid, so perspiration clung to garments and left a film on skin. Imagine, for a moment, ancient style: clothes—washed about as often as the people who wore them—consisted of linen or cotton tunics. They were covered by loose-fitting cloaks made of animal skins, with an extra long flap in front that could be folded up and belted for carrying everything from diaperless babies to small animals—the first pockets. It wasn't as if they ever got aired out, either—the coats often doubled as a sleeping bag, and were likely caked with sweat, dirt, and manure. Everyone wore their hair long; sweaty heads were swathed by veils and headdresses, which protected against the sand and sun. They weren't changed with any regularity, so the olfactory picture is clear: people reeked.[3]

The distinctive smells people emitted due to their diets, work, and lack of bathing are at the crux of one of the Bible's greatest deceptions, the story of Jacob and Esau. Esau was a skillful, if hirsute, hunter, a man of the fields; his twin brother was a quiet, smooth-skinned man who "lived in tents." (Genesis 25:27) Esau was his father's favorite son; Jacob was his mother's. First, Jacob extorts Esau's inheritance—his birthright as the eldest son—from his brother. But Jacob also wants his dying father's blessing. Jacob ponders how to dupe the blind Isaac, and, with Rebecca's help, covers his hairless arms and neck with goatskins. Rebecca realized this would not be enough to fool Isaac in an era when everyone's smell was as instantly identifiable as hairstyles and clothing are today, so Jacob wore Esau's clothing as an olfactory disguise. In the story, Isaac is confused—he recognizes Jacob's voice, but touches the hairy hand of Esau. Jacob, silent, plays his trump card, drawing close enough to receive his father's kiss and, he hopes, his blessing. Isaac "smelled the smell of his garments, and blessed him, and said, 'Ah, the smell of my son is like the smell of a field that the Lord has blessed. May God give you the dew of heaven, and the fatness of the earth, and plenty of grain and wine.' " (Genesis 27: 27–28)

The Bible notes that God himself had olfactory preferences, and the ancient Israelites took great pains to produce them. To an ancient society both anxious to please God and fearful of his dismay, creating and dispensing smells was a physical assurance to the participants that their worship was acceptable.

The Lord commands Moses to slay a ram, "and turn the whole ram into smoke on the altar; it is a burnt offering to the Lord; it is a pleasing odor, an offering by fire to the Lord." (Exodus 29: 18) God also dictates a recipe containing myrrh, cinnamon, and olive oil, to be used for consecrating Aaron and his sons into the priesthood. He instructs Moses to anoint everything inside the meeting tent, from the sanctuary to an altar, for burnt offerings. The sanctity of the perfume was enormous: it was to be used by the priests for the consecration of objects and new priests only, and it rendered holy whatever it touched. Anyone who dared copy the oil, or use it in any other manner, risked estrangement from God and the other Israelites.

Though perfume was an extravagance only the wealthy could afford, fragrant people—likely because they were so uncommon—were the romantic ideal. In the Song of Songs, the scents described by the young couple are metaphorical for their unconsummated love. The boy compares his bride to an aromatic garden—one he longs to enter: "Your lips distill nectar, my bride; honey and milk are under your tongue; the scent of your garments is like the scent of Lebanon. A garden locked is my sister, my bride, a garden locked, a fountain sealed. Your channel is an orchard of pomegranates with all choicest fruits, henna with nard, nard and saffron, calamus and cinnamon, with all trees of frankincense, myrrh and aloes, with all chief spices—a garden fountain, a well of living water, and flowing streams from Lebanon." (Song of Songs 4: 11–15)[4]

Decadence and Decay

Fragrance also infused three pillars of ancient Greece: trade, mythology, and art. One Greek word *aromata,* describes a world of scent: perfume, spices, incense, even aromatic medicine—the first aromatherapy. The Greeks learned of perfumes from the Egyptians and Babylonians and began their own perfume industry; by the seventh century BCE, Corinth was renowned for its intricate terra cotta perfume jars. As in Egypt and Israel, incense accompanied prayer. The gods themselves were said to emit perfume: Homer said that Zeus is "wreathed in a fragrant cloud." Euripides wrote that in heaven, "streams flow with ambrosia."

Sanitation on earth was another matter. People used lidded chamberpots for wastes, and spread them in the fields for fertilizer. They were dumped directly into the streets, so the great cities of Athens, Rhodes, and Thessaloniki (and indeed, cities everywhere) stank of excrement. The poor conditions fostered diseases such as cholera and typhus, and the mortality rate was high. While the link between illness and waste would not be made for millennia, Hippocrates believed that one way to preserve good health was to avoid bad odors. He advanced the Aristotelian notion that air, when combined with the wrong qualities (variations in temperature, humidity, or consistency), was responsible for disease. Noxious air, or *miasma,* had the potential to influence the physical and mental health of all living things.

The cities may have stunk, but people didn't. Homer describes baths among deities and citizens alike; in the cities, there were many public baths. Treatments ranged from dunks in hot water to hot air "baths," or *laconica,* from which the word "laconic" is derived.[5] In Athens, scents were so popular that perfume shops evolved as a counterpart to taverns: people gathered there to buy fragrance, hear news, and trade gossip.

When Alexander the Great conquered Egypt and Persia, he was taken by the vast array of fragrant plants, and sent seeds and saplings back to his teacher Theophrastus in Athens. Theophrastus, in turn, created a botanical garden and wrote the first Western book on smell, *Concerning Odors*. Alexander died in 323 BCE, probably of typhus, and was cremated on a pyre of frankincense and myrrh. His empire crumbled at his death, and Cleopatra, the last of the Ptolemaic rulers, dreamt of uniting Egypt with an ascendant Rome. When her lover Mark Antony entered the port of Alexandria, she is said to have greeted him on a barge with sails that were drenched with fragrance.

The Romans, of course, were notorious for their prodigious appetites, and their indulgence with scent was no exception. In the early days of the empire, in the second century BCE, bathhouses, or *balnere,* dotted neighborhoods throughout the city. The city of Rome itself had eleven public baths and more than 850 private ones. Some were able to serve up to 2,000 people. Inside, fragrant oils and creams were stored in vats, and incense burned continuously. Indeed, the word perfume comes from the Latin *per fumus,* "through smoke," to describe the pleasant smells that drifted through the air when incense was burned.

The Romans were so dedicated to their ablutions that the bathhouse became a central focus of Roman life: people gathered, ate, and even held political discussions there. There were hot water baths, cold water baths, and sweat rooms—even the first public toilets. Seats carved into semicircular blocks of marble were constructed over flowing channels of water that flushed waste to nearby rivers. People wiped themselves with sponges affixed to sticks, left in the latrines for common use.

Since soap was only used for cleaning clothes, hygiene was merely a by-product of the bathhouse "culture."[6] People visited the baths to socialize and relax—even prisoners were taken to the baths regularly. The Romans used fragrance both in and outside the bathhouses. Galen and Dioscorides, two of the era's greatest physicians, investigated the curative powers of aroma; Pliny discussed them in *Natural History.*

The Romans, like the Greeks, thought that stench, or foul air—*malaria*—was responsible for spreading disease. When epidemics struck, they looked to the lethal gases they believed stemmed from the fires and miasmas that lay deep in the earth's core. The philosopher Seneca faulted the sulfuric air released in an earthquake for the deaths of 600 sheep.[7] The poet Lucretius blamed a "deadly breeze" for the plague that swept through Athens from lower Egypt. In the book-length poem *On the Nature of Things,* written in 50 BCE, he described a tree so putrid it could "kill a man outright/by fetid odor of its very flower."

Not only foul air was harmful. Body odors were also offensive, and revealed the social cleavage inherent in Roman society. The fixation with smelling good, of course, excluded those who could afford neither perfume nor hours at the baths—slaves. Their physically demanding chores, from toiling in hot wheat fields to lugging impossibly large temple cornerstones, was no doubt manifest in their body odor. Smelling good, in other words, was a clear sign of status in ancient society.

It was also linked to intelligence. Indeed, knowledge and

wisdom were often described by sensory terms.[8] The Latin word *sapidus,* meaning both "pleasant to the taste," and "prudent" evolved to the French *savoir,* and *saveur;* the Spanish *saber* and *sabor* and the Italian *sapere* and *sapore*—"to know" and "to taste," respectively.

Consider a scene from *Pseudolus,* a play by Plautus:

Pseudolus: But about that slave who's just come from Carytus, is he pretty sharp?
Charinus (holding his nose with a meaning wink): Well, he's pretty sharp under the armpits.
Pseudolus: The fellow ought to wear long sleeves. How's his wit: pretty pungent?
Charinus: Oh, yes, sharp as vinegar.[9]

Smell and the Church

Yet for those who *chose* not to surround themselves with fragrance, there were worse outcomes than ridicule: early Christians who refused to light incense before images of the emperor were put to death. Once Christianity began to take hold in the fourth century, Church leaders called incense "food for demons" and urged the new converts to differentiate themselves from their hedonistic forebears.

Just as the Egyptians, Greeks, and Romans used incense to please their gods, the new religious followers believed that man, in his unwashed state, layered with perspiration, dust, grime, smoke, food, blood, was symbolic of devotion. Denial of earthly pleasures only improved one's status within the Church. In his letters, St. Jerome, the third-century ascetic, wrote admiringly of the martyrs Paula and Melanium:

> Was I ever attracted by silk dresses, flashing jewels, painted faces, display of gold? No other matron in Rome could dominate my mind but one who mourned and fasted, who was squalid with dirt, almost blinded by weeping. . . . Had they frequented the baths, or chosen to use perfumes, or taken advantage of their wealth and position as widows for extravagance and self-indulgence, they would have been called "Madam" and "Saint." As it is they wish to appear beautiful in sackcloth and ashes, and to go down to the fires of hell with fastings and filth![10]

But Church officials fought a losing battle. Early missionaries soon discovered it was far simpler to adapt pagan customs than it was to prohibit them. And while many had abandoned bathing, people clung to rituals—now millennia old—joining fragrance and prayer. By the sixth century, incense and flowers reappeared in church ceremony, serving as symbols of sanctity and grace. Roses, so adored by the emperor Nero (petals were tossed about Rome like confetti during his reign), came to embody Mary; lilies, symbolizing the goddess Juno, came to represent the resurrection. Some scholars say that the origins of the Christmas tree trace in part to Saturnalia, the Roman celebration honoring Saturn, the god of agriculture. For a five-day period beginning on December 17, revelers exchanged gifts, paraded the streets with candles, and decorated their homes with laurel, cedar, and cypress boughs. The festival was so raucous that the elite usually fled the city for the countryside, and the Church outlawed it in 375. But many continued the practice despite the ban. Eventually, Christmas supplanted Saturnalia and winter solstice festivals elsewhere in Europe. But the cutting of evergreens, whether in thanks to Saturn or in homage to Christ, remained a pungent symbol of immortality.

Fragrance Blooms

As the practice of toilette fell into disrepute in the West, it flourished in the East. After Rome fell, Arab scholars carried on many of the scientific achievements of classical Rome, particularly in pharmacy, chemistry, and distillery—crucial, of course, for perfumes as we would recognize them today. The prophet Mohammed (570–632) named "women, children, and perfume" as his top three earthly loves. By the seventh century, the religion that emerged from his beliefs helped also to spawn a quest for more sources of fragrance. The cities that cropped up along the spice route established special inns, caravansaries, for those who carried the precious cargoes of clove, cinnamon, pepper, and rose water. Incense was burned continuously, in tents, in homes, at weddings, and at funerals. Even prayer beads, fashioned of rose petals and resins, released scent when touched.

Ibn-Sina (980–1037), the Arab alchemist, astronomer,

philosopher, and physician, is often credited with discovering distillation (in fact, Avicenna, as he is known in the West, was simply improving on methods that traced to ancient Mesopotamia). He used oils distilled from herbs to treat diseases from cancer to obesity. One of his one hundred books was devoted entirely to the use of roses, which included curing headaches and "fortifying the senses."

In contrast to the early Christians, cleanliness and fragrance were crucial to Islamic worship. (The Koran calls for followers to wash their hands, feet, and faces before the five calls to prayer. It also recommends showering at least weekly, particularly before the Muslim Sabbath, clipping fingernails, shaving pubic hair, plucking armpit hair, and keeping one's mouth "fresh.") Mortar for early Arab mosques was often mixed with musk, so that worship could be infused with scent during the hottest part of the day.

Malodors were especially inauspicious in Muslim society, and linked to evil. In Morocco, for example, flatulence was so much feared because it was thought to blind—or even kill—the angels who resided inside mosques. (No doubt the concern was linked to a diet heavy in beans and grains, but such beliefs abounded even until modern times. So did acts of contrition.) If a person passed wind *outside* a mosque, the action was so connected with demons that people marked spots where it occurred with small piles of stones, as if to entrap the evil. And among the Berber tribes of Morocco, the notion of breaking wind was so shameful and offensive that suicide, in consequence, was not uncommon.[11]

By the ninth century, Muslim traders had reached China, and soon its favorite scents—orange, camphor, and musk (an aromatic substance taken from the gland of an Asian male musk deer)—were incorporated into Islam as well. The Chinese had their own beliefs about fragrance, and classified it into six basic categories, depending on the moods they were thought to inspire—the first aromatherapy. During the invasion of India in the 1500s, one Mughal emperor, Jahangir, built a special garden for his favorite wife, filling it with cedar, roses, and carnations. He called it "Shalimar," or "abode of love," a name Guerlain would use for its great perfume some four centuries later.

But it was the Japanese who advanced perfumery to a fine art. Though incense didn't arrive there until the fifth century, the Japanese soon developed ways to beautify it. Mixing aromatic herbal pastes with seaweed, charcoal, and salt, they pressed them into little cones and figures that were burned on a bed of ashes. Soon, it was used throughout society: Buddhists burned incense in temples, and artisans created boxes inlaid with gold and mother-of-pearl for carrying and storing it. The appreciation of incense even evolved into a game in which participants sniffed unlabeled incense samples to match by scent. The aromatic gums and resins used to make Japanese incense sticks were of such high quality that soon enough, the rate at which it burned became a way to measure time, and by the eighteenth century, geishas were able to tally their charges according to the number of incense cones they used per client.

The Stench of Death

From the Middle Ages to the nineteenth century, the stench of Europe was unimaginably wretched. Foul odors emanated from London to Budapest, worsening as cities grew. The streets were sewers, full of rotting garbage, manure, and diseased and dying horses, donkeys, dogs, and cats. Human waste was dumped into communal cesspits that were cleaned out at quarterly intervals, or from chamberpots directly onto the street. (Scholars believe that the British term "loo" is a corruption of the French *"Gardez l'eau,"* "Look out for the water!") Some cities had privies, separate toilet facilities where users squatted on rotten floorboards—often at risk of falling through them. (Decrepit planks were said to have given way on the Holy Roman Emperor Frederick I in the year 1184.) When the Italian explorer Marco Polo visited the Chinese city of Hangzhou in the thirteenth century, he was shocked to find the sewer system emptying waste into the ocean. He reported that it made the city's air "very wholesome."

Europe's filth, of course, set the stage for the most disastrous epidemic in history: the Black Plague. From 1347 to 1351, the disease, named for the bulging black welts that appeared on the bodies of the afflicted, killed more than 20 million people. Almost certainly the disease was the bubonic plague, transmitted to humans from fleas via black rats that fed on rotting detritus scattered through the streets. Fleas infected with the bacillus, *Yersinia pestis,* fed on the blood of warm rats. When the fleas sucked blood from rat skin, the proboscis would inject the bacillus back into the rat's bloodstream. Once the germ reached the animal's brain, it would convulse and die, and the fleas, highly sensitive to temperature changes, would jump off the dead rat's cold body. Because rats and people lived within such close proximity, the next closest host was a warm human. Within days of a bite from an infected flea, victims suffered fierce headaches, weakness, high fevers, delirium, and uncontrollable coughing. Their tongues turned white, and purple pustules, a sign of internal bleeding, appeared where the flea had bit.

The disease, medical historians say, most likely originated in China and then traveled west to Tashkent. In the port town of Caffa, on the Black Sea, Genovese traders warring with the local Tatars withdrew behind citadels. In what scientists believe was the first instance of biological warfare, the Tatars tossed rotting bodies of plague victims over the fortress walls. The newly infected sailors fled to Sicily, where the disease spread instantly through crowded cities. Authorities tried to cordon off cities as word of the epidemic spread, but it was too late: the sick began to stumble, delirious, about the streets, emitting a terrible stench—everything about them reeked. "Their sweat, excrement, spittle, breath (was) so fetid as to be overpowering. Urine (was) turbid, thick, black or red. . . ."[12]

Theories on the cause of the disease abounded. Some thought it was triggered by the alignment of the planets; others argued that it was the fumes released during earthquakes. Some blamed Jews and lepers, who were flogged and burned alive. Most, though, were convinced that the afflicted had earned God's wrath as a punishment for their sins. The rank odor of the diseased could represent one thing only: the devil, who in medieval minds reigned over a "bog of stench." Simple reasoning gave man but one armament: resisting the disease by avoiding all smells, or with odors fiercer than even the stink of the plague itself.

Officials in the town of Orvieto decreed that men must avoid "fleshly lust and putrefaction with stinking women." Elsewhere, authorities advised against opening windows, so that bad odors would remain outdoors; indoors, people were to stoke cooking fires indefinitely. Pope Clement, in fact, was said to have sat for weeks in Avignon between two bonfires. Other measures were even more desperate. Some people lingered in latrines, breathing in the stench, and smeared themselves with the excrement and urine of healthy people. Others swallowed pus from the boils of plague victims.

Doctors took to covering themselves for protection from head to toe: they wore long leather or woolen gowns, gloves that reached to their elbows, and beak-like masks filled with herbs thought to ward off the disease. They recommended that people sniff nosegays and pomanders made with aromatic plants believed to have antiseptic qualities: lavender, myrrh, and rosemary, among others. Indeed, so many bodies were stacked up in parish courtyards and on the outskirts of cities and towns that they presented nearly as big a challenge as the disease.

As the disease ravaged city after city, it provoked wives to abandon husbands, and parents to flee their children at the first hint of symptoms. Marchione di Coppo Stefani, a Florentine chronicler who survived the outbreak, wrote that the afflicted languished and died alone. "They remained in their beds until they stank. And the neighbors, if there were any, having smelled

The Doctor's Protective Suit

The doctor's robe. *The long woolen robes and strange-looking beak of the doctor's costume were thought to protect as an olfactory shield for doctors battling the plague. The beak was filled with herbs such as lavender and rosemary. Ironically, doctors complained of the fleas, which nested in its folds, helping to spread the disease even further. Illustration from* Historiarum anatomicarum medicarum *(1661), by Thomas Bartholin.*

the stench, placed them in a shroud and sent them for burial." In the nearby town of Pistoia, authorities decreed that families must bury their dead in wooden caskets, nailed shut, two-and-a-half arms deep "in order to avoid the foul stench which the bodies of the dead give off."[13]

Discovery—and Olfactory Shields

The plague continued to erupt across the continent until the early seventeenth century. The Church took extraordinary measures to protect the public as well, consecrating new saints as protectors (such as St. Roch of Montpellier), shrines, and orders of monks as interceptors of the plague. Slowly, public health measures gained currency—Milan, for example, suffered only a 15 percent death rate after it locked up the homes of any diseased person, leaving all family members to die. Indeed, Church authorities began to redouble civic efforts to halt the spread of disease, ordering churches shut down, processions banned, and shrines closed.

People avoided bathing on the theory that exposing one's skin to the air would make one vulnerable: Queen Isabella of Castile noted with pride that she had had only two baths in her life—at her birth and before she got married. But while she herself stank, she cared enough about Spain's edge in the spice trade to dispatch Christopher Columbus to "India." While the *Niña, Pinta,* and *Santa Maria* never reached Asia, the trips to the New World did yield exotic foods and riches. The voyage to deliver them to Spain, however, was an olfactory abomination: no amount of cocoa, peppers, and vanilla beans could mask the stench of the ships, which reeked of vomit, excrement, maggots; mold that had ravaged food supplies, and sopping woolens that never quite dried.

And yet the introduction of these "new" goods, as well as those from the Middle East, ushered in new smells, tastes, and habits. Islamic officials in Constantinople and Cairo tried to ban coffee once it arrived from Ethiopia in the 1510s; they believed its aroma and consumption would distract worshippers from prayer. Still, coffeehouses stayed open, and travelers to the Orient found the drink so tantalizing that they brought it to Western Europe. A French explorer, meanwhile, introduced tobacco seeds to his native land from Brazil. By the 1570s, it was prescribed as "medicine," and smoke enveloped the French and British courts. The Dutch distilled juniper berries into gin in 1650; it quickly gained ground as a popular drink.

Through it all, though, the specter of the plague hovered in the minds—and noses—of an entire continent. Tobacco, coffee, and gin were all touted as prophylactics. People scattered rushes of rosemary and lavender on their floors, and brides carried chives in their wedding bouquets. Meanwhile, though, the reeking of the piled-up dead and the shallow grave pits endured. Stench *was* disease.

In 1721, once again, the plague spread west across Europe, decimating Provence. The Dutch imposed a strict quarantine on all shipping from the East. The port of Rotterdam was choked with smoke and the stink of scorched wood, grain, and fabric. The English government imposed an embargo on all goods from the Mediterranean, infuriating merchants eager to trade in the exotic new commodities.

It was during this time that the writer Daniel Defoe published a semifictional account of the year 1665, when the epidemic struck London. Memories of the year had grown dim, but Defoe drew on medical accounts and diaries to write *A Journal of the Plague Year.* Throughout the book, the protagonist, a saddler, evokes the reek of London and its inhabitants. The calamity, Defoe wrote, was spread by "the breath, or by the sweat, or by the stench of the sores of the sick persons, or some other way, perhaps, beyond even the reach of the physicians themselves . . . (the effluvia) affected the sound who came within certain distances of the sick, immediately penetrating the vital parts of the said sound persons, putting their blood into an immediate ferment, and agitating their spirits to that degree which it was found they were agitated; and so those newly infected persons communicated it in the same manner to others."[14]

Those who somehow escaped the plague were both feared and admired. A gravedigger and his wife, a nurse, had been surrounded by the dying and the dead for more than twenty years, with never so much as a "day of distemper." Their immunity was olfactory: "He never used any preservative against the infection, other than holding garlick and rue in his mouth, and smoaking tobacco. . . . And his wife's remedy was washing her head in vinegar and sprinkling her head-clothes so with vinegar as to keep them always moist, and if the smell of any of those she waited on was more than ordinary offensive, she snuffed vinegar up her nose and sprinkled vinegar upon her head-clothes, and held a handkerchief wetted with vinegar to her mouth."[15]

There was no telling who among the healthy might be sick tomorrow, and whose stink, whether caused by perspiration, filth, or clothes that absorbed the hazy smog of medieval and Renaissance cities, might be contaminating others. The plague generated such immense anxiety about the odors of others that it prompted people to undertake smelly rituals designed to shield themselves from the noxious, uncertain air around them. From the gravedigger's garlic talisman to smoking the tobacco brought from the New World, people enclosed themselves in what some anthropologists now call "private olfactory bubbles."[16]

A Cleaned Up Continent

In Italy, knowledge about contagion spurred public health measures such as quarantine (the word derives from the Latin *quadraginta,* forty days), placed on people and goods coming in and out of cities. This allowed the trading ports of Venice and Genoa to resume trade gradually with the East. Florence, meanwhile, flourished as a center of thought, art, and science. Gradually, these trends popularized cleanliness and fragrance, which had retained its predominance outside of Europe. People began to use imported sandalwood and rose soaps to wash themselves, and bathing returned to vogue. Venice developed a perfume industry of its own, producing scented gloves, stockings, and shoes—even aromatic coins. Cinnamon, nutmeg, cloves, pepper, and ginger added new flavors to drinks and food (and helped conceal decay). Consider the odor of hands alone: cutlery wasn't commonplace before the seventeenth century, so people ate fatty meat, sticky rice, and greasy chicken with their hands (this is one reason why glassware had coarse knobs—it gave diners "traction"). In Italy, noblemen and women resolved this problem by rinsing their hands with bowls of rose water that were set by each plate.

While Italians gradually cleaned up, elsewhere, people tried to cover their lack of hygiene with fragrance. Caterina de' Medici, who married Henry II in 1533, brought the sophistication of her native Florence with her to Paris. Her court included her personal perfumer and alchemist, and together they helped transform a society choking in medieval stink. Caterina chose Grasse, the Provence village, as the perfect site for the herbs and flowers that would go into making her personal perfumes. Yet for all its pleasant scents—it became (and remains) a center for growing lavender, verbena, and bergamot—it was also central to the leather industry. So the sweet wafting of flowers in bloom was juxtaposed with the acrid, foul odor of sixteenth-century tanneries. Soon, perfume was used for, and on, everything in all of France, from hair to gloves to shoes. Even a guild to protect perfumery workers was established in 1656.

By the time Louis XV was crowned in the eighteenth century, perfume use in France was at an apex. Louis himself demanded a different fragrance for his chambers daily. His mistress, Madame de Pompadour, ordered new scents created on whims. She helped to propagate even more uses for fragrance by dousing it on the elaborate wigs she helped make famous, as well as on handheld fans and even furniture. *Potpourri,* a dried blend of rose petals, salt, and spices, became fashionable in parlors and boudoirs alike (before the concoction is fully dry, it is so cloying and strong it earned the mix its name—literally, "rotten pot"). Fragrance also doubled as jewelry: pomanders, scented balls made of ambergris, spices, wine, and honey that dangled from belts or around necks (*pome d'ambre,* literally, "apple of amber" in Middle French).

And while Versailles was known as *"la cour parfumée,"* "the perfumed court," beneath those scents, eighteenth-century

France smelled awful. As writer Patrick Suskind describes pre-Revolution Paris,

> The rivers stank, the marketplaces stank, the churches stank, it stank beneath the bridges and in the palaces. The peasant stank as did the priest, the apprentice as did his master's wife, the whole of the aristocracy stank, even the king himself stank, stank like a rank lion, and the queen like an old goat, summer and winter. For in the eighteenth century there was nothing to hinder bacteria busy at decomposition, and so there was no human activity, either constructive or destructive, no manifestation of germinating or decaying life that was not accompanied by stench.[17]

Indeed, the French historian Alain Corbin writes that in the eighteenth century, smells were so ferocious that much of the nation could "hardly breathe." From the reek of the streets to the iris-scented breath of the bourgeoisie, smells overpowered—and obsessed—France. Scrupulous attention was paid to body odors as portents of disease and keys to seduction. The French army, ravaged by a "fever" outside of Nice in 1799, emitted an "odor similar to phosporous gas in combustion."[18] But other body odors were alluring. A menstruating woman was "conveying the vitality of nature . . . making an appeal for fertilization, and dispersing seductive effluvia."[19] In a famous letter, Napoleon appealed to Josephine "not to wash" for the two weeks prior to his return to Paris. During their tempestuous union, the couple indulged their passion for perfume, ordering gallons of it monthly. Napoleon liked violets; Josephine was partial to the strong, lingering scent of musk. When Napoleon was granted an annulment, Josephine doused the chambers of his new bride, Maria Louisa, with vats of musk, knowing the scent would endure even though her marriage hadn't.

From Chaos to the Cult of Clean: The Smell of America

From the first writings of Columbus to the dispatches of de Tocqueville, the people of America projected a rugged image. While American Indians had frequent baths and herb-infused steam tents, the settlers who arrived were far less fastidious. The Puritans outlawed bathing upon their arrival; they issued jail terms for those caught soaking. Indeed, by the 1850s, a quarter of all New Englanders didn't bathe even once a year.[20] Bodies and clothes were filthy; animal waste littered the countryside, and rats infested sewage-strewn cities. Men spat their chewing tobacco on the floors of taverns, pubs, even the sitting rooms of their own homes. Standards weren't altered until epidemics of cholera, combined with the religious revival of the nineteenth century, swept much of the nation. Like the Methodists in England, who had heard from their founder John Wesley that "cleanliness was . . . indeed, next to godliness," Americans, both rural and urban, began bathing on Saturdays in preparation for church the next day. But preparations for the bath—boiling gallons of water and pouring it into a tin tub—were as dangerous as they were cumbersome.[21]

During the Civil War, the diarrhea, cholera, and typhus spawned by filth killed more soldiers than battle. Because cleaning was seen as "women's work," most young men, away from their mothers, sisters, or wives for the first time, had no idea how to save themselves—or their surroundings—from squalor. ". . . They behaved as boys, whooping it up in the streets with their bugles and drums, getting drunk . . . neglecting to wash and 'change their underwear for weeks at a time.' "[22] Latrines were next to cooking areas, bedding was infested with lice and rodents, and clothing and hair were filthy. Conditions were so wretched that President Lincoln appointed a "Sanitation Commission" for the Union Army, headed by Frederick Law Olmsted, the creator of Manhattan's Central Park. Nurses, influenced by the experiences of Florence Nightingale during the Crimean War, were enlisted to help "scrub soldiers." The writer Louisa May Alcott was among the volunteers. Overwhelmed by the magnitude of filth, Alcott took to guarding herself—and her nostrils—with a bottle of lavender water, which she sniffed as an antidote to "the vilest odors that ever assaulted the human nose."[23]

For years to follow, disease hovered over America's growing cities. As the Industrial Revolution transformed the country from a hardscrabble colonial outpost to an economic powerhouse, the urban population grew from 1.8 million in 1840 to 54 million in 1920. Sewer systems and refuse removal, such as they were, were overwhelmed by factory wastes and mounds of horse manure.

By the turn of the century, New York had emerged as the country's dominant financial and cultural power, but was frequently crippled by outbreaks of typhoid, hookworm, and cholera, all traceable to filth. With its economic viability and physical health under constant siege, the city embarked on a widespread plan to reform public health and improve sanitation. In 1895, it appointed Colonel George E. Waring, Jr., a sanitary engineer, as the city's street-cleaning commissioner. Waring believed that community participation was essential in his goal to clean New York, where litter and sewage from the city's rising pool of newcomers sometimes obstructed whole blocks.

He enlisted the help of "juvenile street cleaning leagues," comprised largely of immigrant children, to help bring the public in line with modern notions of sanitation. Each participant took a civic pledge and compiled weekly progress reports of their respective streets; based on their activity, the children were assigned ranks ranging from helpers and foremen to superintendents. In sections of the city where English was a foreign language, the street-cleaning leagues became an important link to a new, Americanized life—and provided whole families with new ideas of health and hygiene. "Cleanliness," Waring said again and again, "is catching."

In time, cleanliness did catch hold in the national psyche,

shifting from a method of preventing disease to a way of becoming—and being—truly American. By the mid-twentieth century, the nation could boast near-universal plumbing, vacuum cleaners, as well as a plethora of washing powders, shampoos, and toothpaste. With disease now beaten back by vaccines, antibiotics, and cleaner cities and homes, Americans had a new reason to scrub: status. As in ancient Rome, a dirty, smelly body in America now carried serious social implications. Odors—or the lack of them—became a way of defining class.

But just as it became possible to live in cleaner cities and homes, and to wear cleaner clothes on cleaner bodies, the new standards, paradoxically, created a new set of circumstances. Because it was easier to clean and be clean, Americans began doing so obsessively. As Americans strove to conform in the immediate postwar years, advertisers played upon social fears of rejection and spinsterhood by hawking products designed to wipe out embarrassing odors. Newfangled contraptions such as the garbage disposal, a "must-have" item for the new class of suburban homeowners in the 1950s, wiped out kitchen smells, along with flies and rodents. Dial soap could help a person stay clean "hour after hour." The "right" appearance, writes historian Suellen Hoy, "became everything, more important than personality or character in shaping one's fate." Statistics published in *Newsweek* in 1958 reveal this as well as anything: by that year, Americans spent $200 million on products that made them smell better, taking more than 500 million baths a week.[24]

By 2000, Americans spent nearly $10 billion on products to both subtract their odors and add allure, and they took an average of 260 million showers a day. In a mere half-century, America had traversed a terrain of stench and disease to earn a dubious honor that endures today: the most odor-free place on earth.

three

The History of Science (As Told Through the Nose)

The quest to understand the external nose proceeded even more slowly than the attempt to diminish stink. Throughout history, scientists trying to figure out the human body were drawn to the nose, which they believed held clues to the secrets of the brain, the lungs, and even the soul. But until the mid-twentieth century, scholars were ignorant of how the body worked in general, let alone the intricate specifics of the nose. An examination of the nose in science turns up, by today's standards, one absurd notion after the next. Its appearance and host of functions lent it a predominant role in medical fads from physiognomy to reflex irritation theory, and helped to bolster biases about women, looks, and ethnic groups. Each era, in fact, seemed to get the nose it sought: Eighteenth-century phrenologists focused on the size and shape of the nose as a key to personality and intelligence; Freud and Victorian doctors linked the nose to everything from painful cramps to masturbation. As recently as the 1950s, American physicians used ordinary allergies and congestion as proof of psychological maladjustment.

To be fair, the nose held many mysteries. Its dark passageways ferried aromas and breath, as well as a strange, viscous liquid that came and went, often inauspiciously. Galen, the second-century Greek physician, believed the fluid signaled a "purging of the brain" that "percolated" through the base of the skull to the nose. (The notion would endure for 1,500 years.) By the Middle Ages, postnasal drip, hay fever, allergies, and sinus infections were all lumped together as "catarrh," from the Greek word *katarrhous*, "to stream down," and were symptoms of more sinister problems. The sinuses themselves were believed to harbor drawn-in odors that could "taint" the brain, a well for tears, or even evil spirits.

Observers long made connections between the nose and male genitalia: "The truth of a man lies in his nose," Ovid wrote. Virgil wrote of an adulterer who had been "spoil'd of his nose." In the hand-to-hand combat of ancient wars, the nose was a common target for attack. (Perhaps warriors saw it as a way of symbolically castrating their foes.) In one treatise, Renaissance scholar Giovanni Battista della Porta wrote that "the nose is like the rod."

For millennia, doctors struggled to find cures for nasal problems. Papyrus scrolls from ancient Egypt prescribed iris oil for polyps. For colds and stuffy noses, the Chinese gave ephedra (today's Sudafed is a synthetic twin), and the Turks had steam rooms. The efficacy of some remedies remains, even if beliefs about what caused the ailments does not. Deviated septums, for example, were thought to be the result of chronic nose-picking. Polyps were believed to be outgrowths of the brain, bulging through the sinus wall (doctors in Renaissance Italy sliced them off with hot wire "snares"). Elsewhere, others turned to more mundane substances as preventatives for the common cold: in Central Europe, people irrigated their noses with urine; Indians used boric acid; Malians in the Sahara stuffed their nostrils with camel dung.

The sinuses were particularly bewildering. Doctors knew they were connected to the nose, but couldn't find a way to examine them. In order to get a look, doctors sat their patients next to windows, or had them hold their breath while they placed a candle near—or in—the mouth, trying not to singe the flesh. The understanding of the sinuses got an unexpected boost when a patient of seventeenth-century British doctor Nathaniel Highmore stuck a silver hairpin into the gaping hole left by an extracted tooth: "(She) was exceedingly frightened to find it pass, as it did, almost to her eyes. And upon further trial with a small feather stripped of its plume, was so terrified as to consult the Doctor and others about it, imagining nothing less than that it had gone to her brain." Highmore used the opportunity to probe the space, and published a book about his findings, *Corporis Humani Disquisitio Anatomica.*[1]

Physiognomy: You Are Your Nose

The nose was central to the eighteenth-century revival of physiognomy, the pseudoscience of judging a person's character based on the notion that physical beauty displayed moral goodness; coarse features, on the other hand, revealed dishonesty, laziness, and stupidity. The practice dates to Aristotle, but returned to prominence with the 1775 publication of *Essays on Physiognomy Designed to Promote the Knowledge and the Love of Mankind,* by the Swiss theologian Johann Caspar Lavater. The book featured drawings of scholars and artists from Shakespeare to Dürer. Lavater analyzed the nose, forehead, eyes, and chin of his subjects, and gave guidelines for his readers to do the same. A quick glance at anyone, he argued, was all that was needed to understand a person's morality, intelligence, even religiosity. "Physiognomy is the very soul of wisdom, since it elevates the pleasure of intercourse, and whispers to the heart when it is necessary to speak and to be silent, when to forewarn, when to excite; when to reprehend," Lavater wrote in his preface.[2] First published in German, *Essays* appeared in translation throughout the continent well into the nineteenth century.

Its influence was soon felt in art, literature, and social life, both in Europe and the United States. Portraitists rendered George Washington in profile so as to capitalize on his "large, well-shap'd nose," admired by new Americans as evidence of his strength and hawk-like foresight. Napoleon wrote that when he needed a good strategist, "I always choose a man, if possible, with a long nose." And Robert Fitzroy, captain of the HMS *Beagle,* nearly turned Darwin away from its great voyage in 1831 because of his bulbous nose. ". . . On becoming very intimate with Fitzroy, I heard that I had run a very narrow risk of being rejected [as the *Beagle*'s naturalist], on account of the shape of my nose! He was an ardent desciple (sic) of Lavater, and was convinced that he could judge a man's character by the outline of his features; and he doubted wheather (sic) anyone with my nose could possess sufficient energy and determination for the voyage. But I think he was afterwards well-satisfied that my nose had spoken falsely."[3]

Public fascination with physiognomy, and noses in particular, spawned a book devoted entirely to the subject, *Notes on Noses,* in London in 1854. George Jabet, the pen name of Eden Warwick, felt it necessary to justify his subject at the outset: ". . . It might appear prudent, if not altogether necessary, to commence by vindicating the Nose from the charge of being too ridiculous an organ to be seriously discoursed upon. But this ridiculousness is mere prejudice; intrinsically one part of the face is as worthy as another, and we may feel assured that He who gave the *os sublime* to man, did not place, as its foremost and most prominent feature, a ridiculous appendage. . . . We believe that, besides being an ornament to the face, a breathing apparatus, or a convenient handle by which to grasp an impudent fellow, it is an important index to its owner's character. . . ."[4]

The mind, in fact, shaped the nose, Jabet wrote. This theory led to five simple nasal classifications:

I. The Roman, or Aquiline nose, indicated "great decision, considerable Energy, Firmness, and Absence of Refinement."

II. The Greek, or Straight nose, revealed "refinement of character, Love for the fine arts and belles-lettres. . . . It is the highest and most beautiful form the organ can assume."

III. The Cogitative nose, which "gradually widens from below the bridge" indicates "strong powers of thought, given to close and serious Meditation." (In a footnote, Jabet cautioned: "A Nose should never be judged . . . in profile only; but should be examined in front to see whether it partakes of Class III.)

IV. The Jewish or Hawk Nose, very convex, thin, and sharp, denoted "considerable Shrewdness in worldly matters; a deep insight into character, and facility of turning that insight into profitable account."

V and VI. The Snub, or "Celestial Nose" were similar enough to combine in one category. Both revealed "natural weakness, mean, disagreeable disposition, with petty insolence, and diverse other characteristics of conscious weakness. . . . The general poverty of their distinctive character makes it almost impossible to distinguish them. Nevertheless, the Celestial, by virtue of its greater length, (is) decidedly preferable to the Snub . . . and is not without some share of small shrewdness and fox-like common sense."[5]

Jabet discussed feminine noses, which apparently did not fall under such strict divisions, in a separate chapter. On a woman, a Roman nose "mars beauty," and "imparts a masculine energy to the face which is unpleasing (and) opposed to our ideas of woman's softness and gentle temperament." A Greek nose, on the other hand, was highly attractive. The home would surely profit by having such a refined soul at its helm. "It will exhibit itself in her needlework by an artistic arrangement of colours and a poetic choice of subjects; in a neat and elegant attire, in the decoration of her drawing-room, or in the paraphernalia of her boudoir."

Cogitative noses did not much appear on the female sex, as "women rather *feel* than think." Nor did the Jewish nose. "Neither are its indications material to the perfection of the female character. It is the duty of men to relieve women from the cares of commercial life, and to stand between those who would impose upon their credulity."

While Celestial and Snub noses were abhorrent on men, they were, in fact, agreeable on the female face. "We confess a lurking *penchant,* a sort of sneaking affection which we cannot resist, for the latter of these in a woman. It does not command our admiration and respect like the Greek, to which we could bow down as to a goddess, but it makes sad work with our affections." While on a man such a nose represented frailty, a woman, after all, embodied just that: "Weakness in a woman . . . is excusable and rather loveable; while in a man it is detestable. It is a woman's place to be supported, not to support." Jabet acknowledged that in writing about the female nose—"a difficult and nervous subject"—he tread on delicate ground. "We have endeavored, however, to say nothing but the plain truth."

In a final chapter, Jabet discussed "national noses." His thoughts reflected the racist theories advanced by scholars, who along with physiognomy had embraced phrenology, a "science" popular in mid-nineteenth-century Britain that drew conclusions about intelligence and character based on the structure of the skull, jaw formation, and facial angles. "Every nation has a characteristic Nose; and the less advanced the nation is in the civilization, the more general and perceptible is the characteristic form. ". . . the most highly organized and intellectual races possess the highest forms of Noses, and those which are more barbarous and uncivilized possess Noses proportionately Snub and depressed, approaching the form of the snouts of lower animals, which seldom or ever project beyond the jaws."

Jabet's contempt for the noses of non-Anglo-Saxons was unrelenting, but no national nose received more vitriol than that of the new Americans. In stark contrast to their conscientious Puritan forebears, young Americans lacked honesty, integrity, and diligence, Jabet wrote. Its nose, therefore, was "the most unthinking of any Gothic stock." But he did hold out hope. Perhaps one day, he wrote, ". . . It will yet furnish its quota of thinkers to the history of the human mind."[6]

Freud and the Victorians: The Nose as a Sexual Organ

Medicine's view of the nose was a similar mix of surmise and superstition. By the nineteenth century, doctors had begun to fasten on the theory of "reflex action," which held that nervous connections running along the spine joined all the organs of the body together, including the brain. The blood and the nervous system worked in tandem so that a change in one organ could effect sympathetic consequences elsewhere. Conditions in one part of the body easily spread to distant ones: an upset stomach could lead to a troubled mind, and so on.[7]

By the turn of the century, many in mainstream science began to focus on the nose as the culprit in many illnesses, even "slowness." In Great Britain, children who suffered from chronic ear, sinus, and throat infections were thought to be "dull." Like anyone with a bad cold, they breathed through their mouths instead of their noses (this is because adenoids, lymphoid tissue at the top of the palate, become so swollen that they block air coming from the nose to the lungs). Such a habit made a child look dim-witted, doctors thought. One surgeon, William Watson, found that if he removed the adenoids, the children resumed breathing through their noses, and "looked" cleverer almost immediately. As a result, the operation became popular among Victorian doctors and parents alike, who believed the procedure could boost a child's I.Q.

And J.R. Mackenzie, a Baltimore physician, addressed the Baltimore Academy of Medicine regarding the reciprocal relationship between the nose and the genitals. While ancient observers from India to Greece had reported the "evil effects of undue excitation" on sight and hearing, Mackenzie set out to convince his fellow doctors of similar results on the nose. Women were not immune from this most uncomfortable plight. Indeed, they were especially susceptible to such misery on account of menstrual irregularities, which forced blood up to the nose, engorging the tissue. Mackenzie also noted the tendency for young ladies to have "vicarious nasal menstruation"—nosebleeds.

Mackenzie interpreted chronically swollen, runny noses as the body's own punishment for a lack of sexual self-restraint. "Amorous contact" between husband and wife was likely to result in temporarily stuffy nasal passages for both sexes, but more troubling were the chronically swollen noses of those who made a habit of "overstimulating" their genitals. "Repeated and prolonged abuse of the organs may (cause) a constant irritative influence on the turbinate tissue," he warned. (Three pairs of turbinates, membrane-covered cartilage, line the walls of the nose. They are essential for humidifying the sinuses and for protecting them against bacterial and viral infections.) But worst of all were the poor noses of masturbators. "Victims of this vice . . . are constantly subjected to discharge from the nostrils and perversion of the sense, which is simply the outward expression of chronic nasal inflammation."[8]

Mackenzie concluded "that the natural stimulation of the reproductive apparatus . . . when carried beyond its normal physiological bounds, as in coitus or menstruation . . . (is) often the exacting cause of nasal congestion and inflammation." (Embedded among Mackenzie's bizarre theories, this actually had a germ of truth. Decades later, scientists would discover that the nose actually did swell during sexual excitement, but only temporarily. Indeed, the nasal septum is made of erectile tissue like that of the genitals and the nipples. In women, monthly changes in estrogen and progesterone can result in swelling of all extremities, including the nasal tissues, but hormones would not be discovered until the 1930s.)

While Victorian mores and "reflex action theory" made this believable to some—the speech was later published in a prominent medical journal—other scholars found his musings preposterous. Dorothy Reed Mendenhall and Margaret Long, two of the first female medical students at Johns Hopkins School of Medicine, were among those who heard Mackenzie hold forth on his pet topic. "From the start he dragged in the dirtiest stories I have ever heard, read, or imagined," Mendenhall wrote. Though nearly fifty years had passed since that night,

> much of what he said is branded in my mind and still comes up like a decomposing body from the bottom of a pool that is disturbed. . . . Dr. Mackenzie spent most of his hour discussing the cavernous tissue present in the nasal passages and comparing it with corpus spongiosa of the penis. We sat just opposite the speaker and the chairman, so that the flushed, bestial face of Dr. Mackenzie, his sly pleasure in making his nasty points, and I imagine the added fillip of doing his dirt before two young women, was evident. . . . Roars of laughter filled the room behind us at every dragged-in joke. . . . I cried all the way home—hysterically—and Margaret swore. The next few days I stayed at home . . . debating with myself whether I should leave medical school.[9]

But conventional wisdom at the turn of the century was dominated by the belief that sex and sexuality formed the basic core of a person's identity. While the Victorian bourgeois were modest enough to wear stockings even with their bathing suits, their focus on sex and its effects on health was a hallmark of the period.

Masturbation, the "solitary vice" was deeply sinister. Doctors warned that it could permanently maim a child, and urged parents and teachers to keep a watchful eye out for anything suspicious. Some physicians even designed devices aimed at saving boys from temptation: one delivered electric shocks to a sleeping boy's penis should it become erect; another was equipped with bells that enabled alert fathers to rush in if self-control had failed.

Meanwhile, doctors believed that a woman's health centered on her reproductive organs. The uterus and ovaries controlled all parts of the body, they asserted, and were responsible for illnesses from tuberculosis to hysteria. As a "cure" for any range of disorders, doctors in the United States and Britain advocated "leeching" of the vulva (which invariably resulted in "losing" some of the creatures) and injecting liquids such as "linseed tea" and milk directly into the uterus.[10]

Female reproductive organs were inevitably tied to the nose as well. Wilhelm Fliess, a Berlin otolaryngologist, was fascinated by the nose, and claimed that it contained "genital spots." The spots, Fliess believed, grew swollen with "sexual substances" which circulated in twenty-three-day periods for men and twenty-eight-day periods for women; each sex had some of each substance. They coursed through the body in excess if sexual pleasure was not properly achieved—that is to say, if a person masturbated, used condoms, or engaged in coitus interruptus. Fliess gained notoriety for his premise, and published a book on the subject in 1897, *The Relationship Between the Nose and a Woman's Genitalia.*

Fliess had a powerful friend, Sigmund Freud, with whom he corresponded regularly. During the early years of their friendship, Freud became obsessive about his own nose, and filled his letters to Fliess with details of its discharge and inflammation. Fliess also complained of a runny nose, and both men turned to cocaine for treatment; Freud called it a "magical drug," for its mental and physical effects. Apparently, Freud applied his directly to the mucous membrane.[11] Initially, cocaine can reduce inflammation in the sinuses because it dilates blood vessels. But chronic use can irritate, even ulcerate, the nasal passages; cause "rebound" congestion; and perforate the septum.

Fliess treated patients first by applying cocaine locally, then cauterizing tissue inside the nostrils. But only surgery would guarantee permanent relief from "hysteria." Fliess's solution was to remove a portion of bone from the left middle turbinate. This, he said, would permanently cure all female sexual disorders. In a letter to Freud, Fliess wrote: "Women who masturbate are generally dysmenorrheal. They can only be finally cured through an operation on the nose if they truly give up this bad practice."[12]

Fliess's sway over Freud grew paramount in the treatment of one of his early analytical patients, Emma Eckstein, the daughter of prominent Viennese socialists. Like many of Freud's patients, Eckstein suffered from stomach pains and menstrual problems, which Freud traced to excessive masturbation. He summoned Fliess to Vienna in 1895 to perform his signature operation. (Jeffrey Moussaieff Masson describes the ordeal in his book, *The Assault on Truth.)*

Fliess returned to Berlin shortly after the February twenty-fifth surgery. Not only did Eckstein fail to recover, she grew sicker—and even began to stink—as the days passed. On March 4, Freud wrote a panicked letter to Fliess describing her worsening condition:

> . . . the day before yesterday she had a massive hemorrhage, probably as a result of expelling a bone chip the size of a [small coin]; there were two bowlfuls of pus. Today we encountered resistance to irrigation; and since the pain and the visible edema had increased, I let myself be persuaded to call in Gersuny [another doctor]. . . . He explained that the access was considerably narrowed and insufficient for drainage, inserted a drainage tube, and threatened to break it open if that did not stay in. To judge by the smell, all this is most likely correct. Please send me your authoritative advice. I am not looking forward to new surgery on this girl.

Four days later, he sent another dispatch:

> I wrote to you that the swelling and the hemorrhages would not stop, and that suddenly a fetid odor set in, and that there was an obstacle upon irrigation. . . . [Two days later] there still was moderate bleeding from the nose and mouth; the fetid odor was very bad. Rosanes [a third doctor] cleaned the area surrounding the opening, removed

some sticking blood clots, and suddenly he pulled at something like a thread, kept on pulling and before either of us had time to think, at least half a meter of gauze had been removed from the cavity.

Freud seemed more disturbed by his own reaction to the fiasco than he was by Fliess's obvious ineptitude. "At the moment the foreign body came out and everything became clear to me, immediately after which I was confronted by the sight of the patient, I felt sick. After she had been packed, I fled to the next room, drank a bottle of water, and felt miserable. The brave Frau Doktor then brought me a small glass of cognac and I became myself again."

Gersuny performed a second operation, and Freud arranged for Eckstein to recover in a sanitarium. When it appeared as though she was out of danger, Freud acknowledged the "injustice" of the procedure—but not for long. "That this mishap should have happened to you; how will you react to it when you hear about it; what others could make of it; how wrong I was to urge you to operate in a foreign city where you could not follow through on the case; how my intention to do the best for this poor girl was insidiously thwarted and resulted in endangering her life—all this came over me simultaneously. I have worked it through by now. . . . Of course, no one is blaming you, nor would I know anyone who should.[13]

Eckstein nearly bled to death between the second and, implausibly, yet a third operation. They so severely gouged the delicate bones of Eckstein's nose and cheek that they began to cave in, disfiguring a once-pretty face. At first, Freud felt responsible for the disaster, but later blamed it on Eckstein's hemorrhages, which "were hysterical in nature, the result of sexual longing."

Undeterred by Fliess's incompetence, at some point in 1895 he submitted to Fliess's procedures himself. In order to "treat" Freud's sinusitis—likely worsened by cocaine—Fliess cauterized Freud's turbinates to help relieve pain and drain the infected mucus. The surgery appears to have been unsuccessful, but Freud couldn't bring himself to admit it. Instead, he simply applied more cocaine, which he claimed augmented the surgery's "positive" effects. Indeed, he praised Fliess: "in your hands [you hold] the reins of sexuality, which governs all mankind: you can do anything and prevent anything."

In the early half of the twentieth century, Freud's disciples popularized psychoanalysis in the United States. From the 1920s to the 1950s, the nose emerged as a central figure in medical thinking among psychoanalysts and physicians alike. Between the years of 1914 and 1948, dozens of papers on the nose were published in scientific journals ranging from *Laryngyscope* and *Endocrinology* to the *Psychoanalytic Quarterly.*

Early on, doctors readily embraced Fliess's prescriptions for menstrual irregularities. Emil Mayer, the chief of the Ear and Throat Department at Mt. Sinai Hospital in Manhattan, treated women who suffered from cramps by applying cocaine to their nasal passages. He delighted in reporting that invariably within a few minutes of the treatment, a patient's "color came back, her breathing was fine and she went upon her duties instead of to bed as usual."[14]

Intervention on the nose for sexual problems was endorsed by many, but in psychoanalysis, where a man's nose is a powerful "penis substitute" in both dreams and reality, it was interpreted as symbolic castration. In 1929, C.P. Oberdorf, a New York psychoanalyst, wrote of a deeply religious twenty-four-year-old Irish Catholic bachelor who felt so guilty about his sexual desires that he had attempted suicide. The patient, called "Tim O'Brien," longed to have sex in spite of his religious teachings. Yet whenever he was about to have intercourse, he began sniffling so incessantly that he quickly lost his erection. Eventually, he sought treatment for his troubles—the surgical removal of his turbinates. The procedure didn't relieve O'Brien's problem, and it also "flattened" his nose, giving his entire countenance a "tough, coarse appearance" he was convinced would dampen his chances at snagging a suitable bride. Such a nose was sure only to attract "common, low-born girls" such as waitresses and servants.

O'Brien's sexual development revolved around his nose, and Oberdorfer soon determined why. From the time he was a small boy until he was nineteen, he "cuddled" with his mother, his head at her breast, and was devastated at the age of twelve to learn of the reality of procreation. For years he refused to accept a sullied image of his mother, insisting that her purity exceeded that of the nuns themselves. He loved all of her odors, from sweat to gas; they regularly figured in his dreams. And he was both excited and tormented by a recurring fantasy. He longed to shrink enough so that he could enter a woman (especially a nun; the fatter, the better) with his nose by way of her rectum. "Once I was within her body, I could expand to normal size, my legs in her legs, my arms in her arms, moving as she moved—that was the acme of satisfaction."

Oberdorfer called on the established theory of the nose as an "erotogenic zone" in early childhood. "The nose, which in early infancy had assumed the function of the penis, was in turn replaced by the whole face, the head and finally the entire body became the organ for re-entry into, reunion with, the mother-substitute," he wrote. Rather than use the usual sex apparatus, O'Brien instead relied on a penis substitute, the nose, to enter a woman—a mother-substitute. So guilt-ridden was he that he sought "punishment" for the offending organ—surgery, or mutilation of his nose, being the only "logical way out." Oberdorfer concluded: "The nose being the penis, this [surgery] was nothing more or less than a castration. When accomplished, it resulted in all the helplessness of the castrated male body—namely, a female. But even the symbolic castration did not completely remove the need for punishment, nor, for that matter, did it diminish in the least his sexual urge."[15]

The nose was also at the forefront of psychosomatic medicine, which became especially popular in the 1940s after the arrival of the émigré psychoanalyst Franz Alexander. Centers of psychosomatic medicine were founded at hospitals from Massachusetts General in Boston to Columbia Presbyterian in New York, and soon internists and other physicians adhered to the belief that personality type combined with difficult situations to create bodily symptoms. Asthma, for example, was thought to be a conflict between the need for dependency—wheezing was a symbolic cry for mother—and the fear of dependency. Patients with peptic ulcers were believed to equate the need for love with the need for food, much like infants.

Though germ theory was introduced by Pasteur in the mid-nineteenth century, many were unwilling to accept it as medical fact. Leon Saul, a Manhattan psychoanalyst, broached the topic of communicable disease in a paper: "That an infectious agent, a filterable virus, may play an etiological role is apparently established. . . . [But] in certain instances emotional disturbances may play a role." This emotional factor, he wrote, was likely the "most prominent feature in precipitating the symptoms of the 'common cold' and sore throat."

America at Mid-Century: Your Problems Are Your Own Damn Fault

Those twin themes—emotional problems and sexual activity as the culprit in nasal illness—were elaborated on in a series of studies at Cornell University Medical College in Manhattan in the 1940s. The medical treatment of what was delicately termed "women's troubles" had advanced little from the days of Emma Eckstein. If women were treated for their symptoms at all, doctors often approached them radically, with tranquilizers, hysterectomies, or complete mastectomies. Menstruation was referred to as a period of "unwellness," and many of women's physical complaints were dismissed as emotional nattering. Not far behind women on the ladder of medical scorn were immigrants, namely Jews, Irish, and Italians, with African-Americans on the lowest rung.

In the 1940s, three Cornell professors, Thomas Holmes, Stewart Wolf, Harold G. Wolff, and an assistant, Helen Goodell, conducted research to determine the extent to which patients' nasal symptoms—stuffiness, postnasal drip ("vasomotor rhinitis"), sneezing—were related to anxiety, depression, or "hysteria." The experiments were published in a 1950 book, *The Nose: An Experimental Study of Reactions Within the Nose in Human Subjects During Varying Life Experiences.*

The description of the case studies—and what the doctors saw as the "real" problem—leaves little to the imagination: "Vasomotor Rhinitis and Sneezing in Dissatisfied, Frustrated, Resentful, Weepy Woman." "Chronic Disease of the Sinuses and Vasomotor Rhinitis in an Anxious, Dissatisfied, Resentful Woman who Based her Security on her 'Good Looks' and 'Sexual' Assets, and Who Feared She Was Losing Both." "Chronic Rhinitis, Polyposis (35 Operations) and Ultimately Asthma in a Frustrated, Resentful and Defeated Woman."[16] Others were similarly described: "a rejected, insecure Jewess who experienced strong feelings of resentment and nasal obstruction." "Nasal Obstruction in an Insecure, Dependent, Ambitious, Lachrymose Man." Charts and graphs accompany the patient summaries, complete with traumatic histories corresponding to their nasal symptoms: the death of a parent, the loss of a job.

1916	BORN	PARENTS ILL TBC	
1919	AGE 3	PARENTS DIED	
	AGE 1-5	UNWANTED NEGLECTED	
1921	AGE 5	ADOPTED DYNAMIC FAMILY	MANY HEAD COLDS
		1 FOSTER SISTER	
1934	AGE 18	COLLEGE	
1936	AGE 20	DEATH OF DEVOTED FRIEND	
		GRIEF	VASOMOTOR RHINITIS
		FEAR OF BEING DESERTED	
		WEEPING	
1937	AGE 21	FAMILY ARGUMENTS	
		INSECURE	VASOMOTOR RHINITIS
		FEAR OF LOSING FAMILY	
		WEEPING	SUBMUCOUS RESECTION
1938	AGE 22	ENGAGED AND MARRIED	
1940	AGE 24	FIRST PREGNANCY	V. RHINITIS ABSENT
1943	AGE 27	SECOND PREGNANCY	
		HUSBAND ENLISTS	
		WEAK HEART	VASOMOTOR RHINITIS
		FEAR TERROR WEEPING	
JAN. 1944		SON BORN	↓
	AGE 28	DIFFICULT LABOR	
		HUSBAND LEAVES FOR NAVY	
FEB. 1944		CONFLICT WHERE TO LIVE	
		LONELINESS	V. RHINITIS SEVERE
		FEAR AND WEEPING	↓
MARCH 1944		FOUND ALLERGIC TO	
		COFFEE AND CHOCOLATE	
		BREAKING UP HOME	V. RHINITIS UNCHANGED
MAY 1944		HOUSE NEAR HUSBAND	
		COLD FOGGY WEATHER	V. RHINITIS IMPROVED
SUMMER 1944		LIVING WITH HUSBAND AND	
		CHILDREN	
		CHOCOLATE AND COFFEE	(NO V. RHINITIS)
JAN. 1945		HUSBAND ORDERED TO SEA	
	AGE 29	PANIC AND FEAR	V. RHINITIS 12 HOURS
		WEEPING	
SUMMER 1945		BACK IN HOME	WELL

Lift chart illustrating coincidence of situational threats and nasal disturbances in an insecure, anxious woman. The black bars at the right indicate the occurrence, duration, and intensity of troublesome symptoms.

In a foreword to the book, the authors wrote that their aim was to "study the man and his nose at once as a unit, thus integrating these points of view for the better understanding of the human organism." But the methodology of studying that integration seems about as rational as the theory of "genital spots." The authors—who insisted that they avoided the subjects' "discomfort" or "inconvenience," although some were examined daily for as long as a year—seated their patients next to a lamp, and, using a nasal speculum, peered up their noses to chart the level of swelling, amount of secretions, and color of the mucous membrane. The subjects were asked about their "dreams, prevailing attitudes, preoccupations and moods" in order to determine the relationship between their nasal symptoms and "life situations." Before interviews, many were given Amytal, a sedative hypnotic occasionally used in psychotherapy with trauma patients to "access" repressed or unconscious feelings and memories. (Even though Amytal has been referred to as a "truth serum," psychiatrists today say that it does not guarantee honest recall of events any more than any other interview technique.)

The authors said there were clear links between humiliation, rejection, anxiety, and dejection to everything from runny noses to sinus disease. In one case, a forty-year-old itinerant salesman had recurrent stuffiness ever since his wedding fourteen years earlier. The son of an abusive Russian Jewish immigrant father and a "tearful, unhappy mother," the man quit school to help support his ten younger siblings. His adult home life was scarcely happier than the one he had dreamed of abandoning as a child. The couples' sex life was miserable, and the man's wife was a bitter, resentful woman who had been "rendered sterile" by a botched surgery.

"I've been robbed of a compatible marriage," the patient is quoted as saying. "Robbed of a family. Robbed of the opportunity to do the things I wanted to do. . . . My wife held me back . . . She made a damn fool out of me. A jackass. . . . Yesterday I accused her of being 75 percent the cause of my sickness."

Wolf blamed the man's wife, too. He wrote that whenever his hapless patient was reminded of his wife's "lack of devotion," his nose began to swell and run.[17]

The book devotes a separate chapter to the nose and sexual

function, and the doctors cite Dr. Mackenzie liberally. One "observation" involved a twenty-nine-year-old doctor who complained of a persistent runny nose and stuffiness, from which he was freed only by having intercourse with his wife, "a willing partner to the sex act." However, months later, the doctor "tenderly insisted" on lovemaking with his wife only four weeks after she had given birth to their first child. Despite the wife's hesitation—she was afraid of "sustaining injury," the authors wrote—the couple nevertheless proceeded. The husband, aware of his wife's reluctance, "reacted to the situation with feelings of guilt and humiliation." During fondling, he noticed a "moderate increase in obstruction to breathing." During "copulation," obstruction became "complete." The authors, as they do throughout the book, draw large conclusions from minor events. "After ejaculation, the subject was aware of a small amount of secretion in both nasal cavities."[18]

One seventeen-year-old woman had her nose examined several times during her twenty-hour labor and seven-day postpartum hospital stay. The patient had a difficult labor, during which she struggled to breathe; further, she was grappling with the dueling visits of parents and new in-laws, who openly disapproved of the shotgun marriage. The authors, unlike Mackenzie, attributed this woman's nasal problems, as well as those of other female patients, to personal circumstances. They wrote that those who "accepted pregnancy as desirable" sailed through pregnancy without nasal symptoms. Those "to whom pregnancy posed a threat to the maintenance of their precarious security props," whatever those may have been, developed or experienced increased symptoms during gestation.[19]

In a chapter that focused on "physical threats" to the nose, the doctors examined a fifty-eight-year-old man who suffered from a severe ragweed allergy in March, when he was symptom-free. Doctors stuffed ragweed pollen up his nostrils to see if he would "react." (Antihistamines, meanwhile, were synthesized for medical use in the early 1940s by the Swiss researcher Daniel Bovet, who would later win a Nobel prize in medicine.) The patient suffered an acute allergic attack, lasting for more than ninety minutes.[20]

Finally, the doctors attempted to explore the "sinus headache" using eighteen patients and themselves as healthy controls. The experiment consisted of inserting an electronic probe into the nostrils and sinuses and sliding a catheterized balloon into the sinuses and "blowing air through it." Not surprisingly, the doctors discovered that these implements provoked considerable pain, especially in those already suffering from disease.[21] This research, which went on to earn Dr. Wolff renown as a pioneer in the study of headaches, found that the turbinates and ducts of the sinuses were much more pain-sensitive than the lining of the sinuses themselves.

The authors concluded their book with some sweeping thoughts. "Nasal obstruction," they wrote, "though effective in keeping out dust and irritant gases, is less effective against blows or unrequited love. . . . Closing the air passages and increasing the mucous membrane secretions [minimize] the damaging effects of a gas or irritant, but . . . do not protect against a thorn in the foot or make less destructive the hostility of a parent or marital partner. Indeed, they often lead to additional distress."

The procedures the Cornell doctors performed were hardly the only unscientific studies conducted in the name of medicine in the 1940s. Nazi doctors, of course, were convicted of the atrocities they committed on concentration camp prisoners—submerging them in freezing water to determine how long the body could withstand cold; deliberately infecting them with smallpox and malaria; operating on them without anesthesia. And, throughout the Cold War, the U.S. government enrolled unsuspecting patients in "research" designed to test the body's absorption of radiation. But perhaps most notorious was the Tuskegee syphilis study from 1932 to 1972, in Macon County, Alabama. Over those four decades, U.S. government doctors examined and withheld treatment from 399 African-American men infected with the disease. Though penicillin was found to kill the syphilis bacterium in the early 1940s, it was still denied the men, who were given aspirin for their discomfort and told they were being treated for "bad blood."

While the medical establishment had embraced informed consent after the Nuremberg Trials in 1945, the disclosure of the Tuskegee study in 1972 finally made it an unquestioned requirement for experiments involving human subjects.[22] As bizarre as it might seem today, the Cornell research would hardly have raised an eyebrow in its time.

The doctors' theories about the underlying causes of sinus

disease—emotional stress and sexual frustration—reveal more about their own preconceptions and biases. Studies of the nose were particularly open to this sort of misconception, since the technical tools that would allow a true test of the hypothesis had not yet been invented. Because their functions were neither visible nor measurable, the limited data could be manipulated to say whatever was needed.

In recent years, experts in nasal disease have identified the environment and, increasingly, genetics, as the root cause of allergies and sinus problems. And there is some evidence that the Cornell doctors were right about some aspects of their work, albeit for the wrong reasons. Recent studies suggest there is some relationship between psychological health and the immune system. And it is now understood why a woman's nasal passages can swell in pregnancy and at certain points in the menstrual cycle.

Still, the Cornell interlude is one medical science seems determined to forget. Dr. Harold G. Wolff's research and reputation in the cause and treatment of headaches continues to be held in high regard; a postgraduate fellowship at the American Headache Society is named for him. His data on headaches continue to be cited without reference to how they were obtained.[24]

To be sure, medical science made some astounding breakthroughs in the mid-twentieth century: penicillin was isolated for antibiotic use in 1940; the Pap smear was introduced as a routine cancer screen in 1943; DNA was discovered in 1944; smallpox was virtually eliminated worldwide by 1951; and Jonas Salk introduced the polio vaccine in 1952. But the science of the nose involved little more than guesswork, and had as much relation to a cure as throwing spilled salt over your left shoulder does to ensuring good luck. A biomedical revolution was beginning to take place, but the nose was stuck in the Dark Ages, a mysterious, feared, and little-understood organ.

four

Nagasaki Up the Nose

Monel Metal Nasopharyngeal Applicator

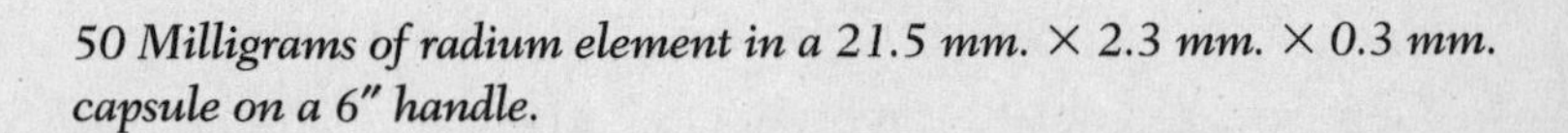

50 Milligrams of radium element in a 21.5 mm. × 2.3 mm. × 0.3 mm. capsule on a 6″ handle.

While doctors at Cornell labored to prove the relationship between the nose and psychological problems, physicians in Baltimore knew its very real links to bona fide ailments such as deafness and sinus disease. They were intent on finding a treatment for upper respiratory problems, and came up with what they were sure was a stroke of brilliance. But little did they imagine that their revolutionary cure—radium administered up the nose to shrink adenoids and tonsils—would mar the lives of millions of patients.

In the early half of the century, hearing loss, especially for children, was a real and present threat; such a disability could marginalize a person for life. And by 1948, chronic upper respiratory infections had left an estimated 4 million Americans under eighteen deaf.[1] Prior to antibiotics and vaccinations for rubella (which, if contracted by a woman in the first trimester of pregnancy, can lead to the blindness and deafness of her child), hearing loss was an entrenched and perplexing problem. Samuel Crowe, the director of the Department of Otolaryngology at Johns Hopkins University in Baltimore, was particularly troubled by the plight of young patients who could hear one month, and barely make out their mother's voice the next.

To prepare for his post at Johns Hopkins, Crowe had trained in Germany, which in the early part of the century was home to some of the most pioneering ear, nose, and throat researchers. In Germany, physicians experimented widely with radium in the shrinkage of animal tumors. By the late 1920s, radium was used to treat polyps in the United States. (Pierre and Marie Curie discovered radium in 1898, and found that it killed human cells. They reasoned that if it destroyed healthy organisms, it could do the same for unwanted tissue as well.)

In treating young patients at a Maryland clinic for the deaf, Crowe wondered if radium might be suitable for their problems. The culprit was swollen adenoids, lymphoid tissue located behind the nose and soft palate, which blocked the Eustachian tube and resulted in hearing loss. While it was possible to remove adenoids surgically, the procedure was risky, painful, and was not always permanent; the tissue sometimes grew back. Radium, Crowe believed, would not only destroy adenoids, it could be administered simply, without anesthesia.

In the early 1940s, Crowe also sought out army and navy doctors to discuss the use of his radium on servicemen. Airmen and submariners were routinely suffering hearing loss and severe ear pain due to the lack of pressurized cabins in aircraft and submarines, and military doctors had been hard pressed to come up with a solution as they lost personnel. (Approximately one-third of all U.S. Air Force aviators were grounded during the war because of their hearing loss, called "aerotitis media.") The navy enthusiastically agreed, and before long Crowe and his protégé, Donald F. Proctor, set to work on designing a device that could deliver the radiation.

In 1943, with the help of the New York–based Radium Chemical Company (which later gained notoriety as one of America's most polluted Superfund sites), Proctor came up with suitable tools: long, wire-thin rods with radium tips, to be inserted up into the nostrils until they reached the swollen tissue. The rods, left in place for eight to twelve minutes, hit the adenoids with at least 2,000 rads of radiation, although some estimates are much higher. Nearby organs such as the brain, thyroid, and pituitary glands also received substantial doses of radiation; estimates range from a few rads to 150. By comparison, cancer patients undergoing radiation to shrink their tumors get some 6,000 rads; an average chest X-ray delivers about .01 rad. Scientists today generally agree that the government limit for occupational radiation exposure—five rads per year to the whole body—is safe.

The standard protocol was to repeat the procedure two

times, at two-week intervals, for a total radiation dosage up to 2 million times that of a typical dental X-ray. The army first called the rod the "Army Irradiator," but later used a less martial moniker, dubbing it the "Aerotitis Media Control Program."

The treatments were hailed as a miracle cure for the servicemen. One study found that 90 percent of those who received them—and more than 8,000 did—were permanently relieved of their ear problems. Before long, their use caught on among civilians, too: by current Centers for Disease Control and Prevention estimates, between 500 thousand and 2.6 million children were treated with the devices between 1948 and 1961. The rods were lauded by the *Saturday Evening Post* in 1948 in an article titled "Will Your Child Be Deaf?" The treatments, known as nasal radium irradiation, or NRI, were a "spectacular" and "remarkable" way to prevent deafness and upper respiratory problems, the article said. The therapeutic "limit" for NRI was ostensibly two courses, but some children received eight or twelve doses, often separated by a week or less. The treatments hurt. Children screamed in pain, and parents were routinely enlisted to hold their charges down. Proctor noted that if very young children could not be kept still, they should be strapped into special chairs that immobilized their heads during treatment. Many complained of a burning sensation that lasted for weeks. The article didn't mention such unpleasantness, however. Rather, it noted that children who had done poorly in school went on to blossom after the radium, both academically and socially.

Radium, in fact, was in use everywhere, and for the most

part, the attitude toward it was relaxed. Nuclear bombs were detonated in the open desert; shoe store customers could slip their feet into X-ray machines to examine their bones, and special X-ray photo labs offered lovers the opportunity to immortalize the intertwined bones of their fingers on film. Surely, a dab of radiation up the nose could hardly be harmful. On the contrary, radiation appeared to be a cure-all: doctors used it not only in X-rays and killing cancer cells but also for treating acne and removing birthmarks.

Nasal radium had not been subjected to rigorous scientific evaluation either. Long after he had been using the procedure widely, between 1948 and 1953, Crowe undertook a study of Baltimore schoolchildren. Its aims were to determine "the feasibility of irradiation of the nasopharynx as a method for controlling hearing impairment in large groups of children . . . (and) to draw conclusions concerning the per capita cost of such an undertaking as a public health measure," he wrote. Further, Crowe added that while the procedure itself was not new, "this is the first adequately controlled experiment of sufficient size for accurate statistical analysis."[2] (Not surprisingly, the study found that the treatments were highly successful.)

Warnings about the overuse of radium in medical treatment first surfaced when Marie Curie, who was awarded two Nobel prizes for her research with the metal, died in 1934 of radium poisoning, which experts traced to her constant exposure. (Pierre was killed in an accident shortly before the couple received their first Nobel, in physics.) By the 1950s, unusual head and neck irradiation was being linked to thyroid cancer. But the medical profession was slow to cut back on radiation use.

Throughout the 1950s, the treatment was as accepted in some circles for a child's nasal allergies and ear infections as Tylenol is for fevers today.

Not everyone was enthusiastic. Two Boston doctors, Laurence Robbins and Milford Schulz, presented a paper at a 1949 meeting of otolaryngologists, stating: "That radiation of any sort should become a routine is unwise; certainly if this does occur, the results should be checked and rechecked at intervals in order to anticipate an untoward effect. Lack of an immediate reaction should not lull to rest the fears for possible latent reaction." The decision to treat a benign condition with such a potent agent as radium should be weighed carefully, they warned, as "unknown dangers may not appear for as long a time as 10, 20, or more years."

Dr. Kenneth Day, a Pittsburgh ENT, also had growing concerns about the overuse of nasal radium, and the same year presented a paper at a different annual meeting of otolaryngologists. "There is no question it has become big business. . . . the use of radium applicators in the nasopharynx has definitely reached the racket level," he said. Some doctors were even using the device "for such alien conditions and symptoms as head colds, tinnitus [ringing in the ears], and chronic cough." Day warned that the use of the radium rods—which could be used successively on an "officeful of patients"—could become a lucrative source of income for the user. "We are rapidly reaching the point where the cure is worse than the disease," he continued.[3]

Still, the radium rods became a routine. Proctor assured parents in a 1960 book that "These treatments have now been given . . . to hundreds of thousands of patients and no instance of damage from irradiation has yet been reported. Nevertheless, both because two series of well planned treatments should suffice, and to eliminate any danger from excess irradiation, the limit should be set at two."[4]

By 1970, nasal radiation was supplanted by antibiotics and, in chronic pediatric cases, drainage tubes were surgically inserted directly into the eardrum. The treatment became less popular as the public realized the dangers of radiation, and eventually faded from sight.

About ten years later, Stewart Farber, a public health researcher, was having coffee with a childhood friend. Michael Krabach, a recreational diver, mentioned that in the late 1960s a colleague had recommended that he see Henry Haines, a doctor from Connecticut, who had a unique treatment—nasal radium—to help divers adjust to water pressure changes while underwater. Krabach (who had so far suffered no ill effects from the treatments) thought little of it until his meeting with Farber. "I was absolutely floored," Farber says. "I thought he was putting me on." Krabach insisted that Haines said the treatment was safe, and had been used on thousands of people for decades.

Farber, fifty-five, a chemist with a degree in public health who had been involved for about a decade in conducting radiation monitoring programs around nuclear plants, saw the "whole sorry episode," as he calls it, through a unique prism: his own training in the body's tolerance for radiation. "I'm the opposite of no-nukes activists," he says. "I've worked to promote the safety of nuclear radiation my whole life." A powerfully built man with brown hair and a trim graying beard, Farber's long résumé includes posts as a senior radiological engineer with the New York Power Authority, and, four months after the Three Mile Island accident, as the assistant manager for nuclear information with the New England Electric System.

Farber began the task of researching Haines's treatments and their repurcussion. Haines, it turned out, had been trained by Crowe, and began work in the mid-1940s at the navy's submarine school in Groton, Connecticut. It didn't take long for Farber to conclude that the whole experience was "totally botched." He began to scour records for studies, which were scarce. One was a doctoral thesis published in 1980 by a Johns Hopkins researcher, Dale Sandler, which found twice the rate of both benign and malignant tumors among those treated with the rods as among a control group. Treated subjects had a higher rate of brain cancer and a nearly ninefold increase in the rate of thyroid disorders such as Graves's disease. Strangely, the women Sandler studied had a lower-than-average rate of breast cancer, but the figures were not statistically significant.

A Dutch paper offered a conflicting view. In 1989, Dr. Peter Verduijn found that the nasal radium had not been linked to a single death. Moreover, there was no incidence of breast cancer among the women who had been treated as children. (There are some possible explanations for the differences. Farber was not persuaded. Dutch children who received the treatments had a 3.5 times lower rate of radium delivered to their adenoids than did the American youngsters. At the time, the two reports were the only significant nasal radium studies.

Farber was appalled at what he was learning. "It wasn't clear that anyone was tracking these people, spread all across the country, and notifying them of what their risks were. You've got people saying, 'Oh, cell phones! Cell phones cause cancer!' And here millions of people had these treatments, using massive doses of radiation. By comparison with any other head and neck cancer risk, it was a Nagasaki up the nose! And who cared? Nobody!"

Outraged, Farber made it his mission to raise public consciousness about the issue. He found a window of opportunity in 1993 when the *Albuquerque Tribune* published a series of articles on government-sponsored experiments in the 1940s in which civilian hospital patients were injected with plutonium. The series won national attention (and earned a Pulitzer prize for the paper). Other stories of radiation experiments emerged, including one in which the testicles of prisoners in Oregon and Washington were irradiated; another involved scientists from Massachusetts Institute of Technology who fed mentally retarded children at a Boston-area school breakfast cereal and milk with radioactive tracers to study the absorption of iron and calcium.

Soon afterward, President Clinton appointed a fourteen-member panel, the Advisory Committee on Human Radiation Experiments (ACHRE), to investigate the Cold War treatments that took place under the auspices of national security. It calculated that children who had received the nasal radium faced a 62 percent higher risk than the normal population of getting brain cancer. These statistics put nasal radium in a special category: its high risk exceeded the committee's stated threshold of one case per thousand for mandatory notification and follow-up.

But the committee decided not to notify patients. After all, those at risk for long-term complications had probably already found them—or were dead. The report said that the brain, head, and neck were most at risk, but that there was no "accepted" screening procedure. If new screening measures were developed, or new information about the risk was discovered, the nasal radium experiments should be reevaluated. Still, Dr. Eli Glatstein, a committee member and the former Chief of Staff of Radiation Oncology at the National Institutes of Health, recommended monitoring military personnel.[5]

Dan Guttman, who served as executive director of the committee, says that finding people where the criteria are met would be a logistical ordeal. Many of the records are missing and incomplete. Logs noting the treatments exist in small hospitals throughout the country, but many lack patients' names. How do you notify such people—and toward what end? Guttman asks. "We had to worry about creating a national panic."

Meanwhile, a Veterans Administration study compared deaths among 1,214 submariners treated with the nasal radium against deaths in a control group of 3,176 randomly selected veterans who were not treated. The report found that those who had had the NRI had a 47 percent higher rate of death from head and neck cancers and a 29 percent higher overall mortality rate than veterans who weren't treated.

Not everyone read the data the same way, and many have clung to the early Dutch data that had shown no serious side effects. In fact, the government has provided no definitive statement for those who received the treatment. On the contrary, the government's actions have been inconsistent at best. The CDC and the Veteran's Administration concluded that further testing was unnecessary among healthy adults. Citing the Verduijn and Sandler studies, they said: "Current studies do not indicate substantial increases in risks . . . among those who received NP (nasopharyngeal) radium treatments."[6]

The CDC has, however, suggested that doctors consider head and neck examinations for patients with a history of nasal radium, and that they ask patients over thirty-five with head and neck complaints about past irradiation, presumably to spot any problems that might be developing.

The CDC has devoted an extensive Web site to nasal radium treatments, in which it urges patients who had nasal radium to notify their doctors. What of those who were toddlers when they received the treatments, or simply don't remember? Dr. Paul Garvey of the CDC shrugs this off, saying that most would recall such an uncomfortable treatment. "We don't feel that this presents a significant problem."

Unfortunately, the CDC may be relying on out-of-date data. Even Verduijn, the Dutch researcher, produced new research showing that the treatments may not have been as benign as his initial findings indicated. In a follow-up study in 1996, he found that those treated with nasal radium had a doubling of overall cancer incidence, with most of the excess due to head and neck cancers.

As Farber puts it, "These people are in a Catch-22." On one hand, the government says the existing studies aren't comprehensive enough to attribute various disorders to the treatments; on the other, it is not willing to pay for a larger, more exhaustive study. "Either way, there are hundreds of thousands—millions—of people out there who may not even know about their risks," Farber says. "What about them? It's severely impeachable science," he says. "The children who received these treatments got zapped at a much higher rate than did the adults, simply by merit of the size of their bodies."

The government eventually took a different view toward veterans. In 1998, Congress required the U.S. Department of Veterans Affairs to provide health care to the estimated 8,000–20,000 veterans with head or neck cancer who had the treatments.

Meanwhile, Farber became the center of the loose network developed by those who had had the treatments, and formed a group, the Radium Experiment Assessment Project, which he runs from his home in Warren, Vermont. More than a thousand people have contacted him in search of information about NRI and its potential risks. Further, in logs he has kept from those who had received the treatment, odd repetitions began to appear, as those patients, now in their forties, fifties, and sixties, complained of tumors, thyroid and immune disorders, brittle teeth, and reproductive problems.

The effects of the radiation appear to be complex. Studies show a lower-than-average incidence of breast cancer among those treated with NRI. Researchers theorize that the radium dispersed by the rods somehow disrupted the function of the pituitary gland to produce its growth, reproductive, and labor-inducing homones. Oncologists, meanwhile, have linked reproductive homones to breast cancer.

Many of the women treated with NRI report an array of similar, strange symptoms. Jan Nelson, a California woman in her mid-fifties, had several treatments for chronic ear infections as a child in Los Angeles. She has had several precancerous nodules removed from her thyroid and suffers from Sjögren's syndrome, an autoimmune disorder that targets the body's moisture-producing glands and causes dryness in the eyes and mouth. Sjögren's is largely hereditary, but Nelson has no family history of the disease. In addition, she suffered bizarre menstrual irregularities that few doctors could puzzle out. Her cycles sometimes lasted for months on end, and while she did have two full-term pregnancies, she went three weeks overdue with both children, and her labor was finally induced. Her doctors told her that her body simply didn't produce a labor-triggering hormone. Rosemary Noel, a Kansas schoolteacher who was treated with radium in Mississippi as a child, suffered from excessive monthly periods until she was thirty-four, when she underwent a hysterectomy. Arlene Lavin, a California woman who also received the treatments, suffered similar problems, leading to two procedures while in her thirties to stop the bleeding.

Most of those who reach Farber seek information on head-and-neck tumors and thyroid irregularities. John Rushton, a computer systems engineer in Pennsylvania who was treated with NRI at a clinic in Elkton, MD, in the mid-1960s is typical of those who have contacted him. Doctors said tissue that had returned after a tonsillectomy and adenoidectomy was exacerbating his asthma, so they recommended nasal radium. He thought nothing of the treatments until Johns Hopkins University Hospital notified him in 1976 as part of a study on increased risk for thyroid cancer among people who had received radiation to the head, neck, or chest. As a Johns Hopkins patient for endocrine problems, his records were already on file. A thyroid scan showed some irregularities, and surgery revealed a disease called Hashimoto's thyroiditis, an autoimmune disorder of the thyroid gland, which doctors said could well be traced to NRI.

In 1995, Rushton found a malignant lump on his neck, later diagnosed as non-Hodgkin's lymphoma. He had six months of chemotherapy and remains cancer-free, but he worries constantly about the disease returning. His oncologist at the University of Pennsylvania thought the disease was likely traceable to the childhood radiation. "I'm not a fearful man, but health problems have dogged me my whole life. I'm not saying they're all tied to NRI, but when even your doctors wonder, you sure do, too," Rushton says. He also suffers from a glandular problem in his leg that doctors link to his lymphoma.

Rushton has read a great deal about the treatments, and follows each developmental blip carefully. He is puzzled by authorities now telling him not to worry about longitudinal studies about NRI, when in 1976 officials at Johns Hopkins had the wherewithal to contact him about his risk for thyroid cancer. "I was just shy of 18 and I've thought about it every day since."

While they cope with dire diagnoses—Graves's disease

here, cancerous thyroid cells there—those in Farber's network also battle rage. On a message board Farber established for those who had the treatments, notes are often darkly despondent. "Still angry after 40 years," reads one. "I hated that dr. . . . [when he died] I felt no sympathy for this miserable piece of trash of a man. I still have nightmares." Another man, just learning of his risks from press reports, writes: "Even as a child I wondered why the doctor who gave me the treatments had to leave the room due to excess radium exposure, when I was the one lying there with the rods up my nose."[7] Many on the board lionize Farber, and there is a floating suggestion that he receive a grant or award—some kind of national recognition—for his efforts. None of it gives Farber any comfort. "I'm totally burnt out," he says. Frustrated, feeling abandoned by the scientific community, and bankrupt—the energy and time he had devoted to the subject left him more than $100,000 in debt—Farber persists, but is resigned to small victories.

Massachusetts alone took firm public action: in 1997, it became the first state in the nation to alert doctors to the risks associated with nasal radiation. The state Department of Public Health's advisory instructs doctors to perform "thorough head and neck examinations" on patients who say they've had the treatment and to report any apparent health effects to the state. It is the strongest step yet taken in response to nasal radiation concerns.

Radiation facts and figures flow from Farber like batting averages from a baseball junkie. Former radiation workers at nuclear weapons facilities whose cancers are proven to be related to excess radiation exposure—in some cases, less than the standard safe limit presented by the CDC, five rads per year—are eligible for $150,000 lump-sum payments approved by Congress in 2001. "Compared to the 2,000 rads or more per nasal radium treatment, that's nothing!" he exclaims.

In 1996, Farber began gathering research for attorneys who were interested in filing suits on behalf of NRI patients, but soon, the case looked hopeless. The Radium Chemical Company, plagued by its own environmental and legal troubles, was defunct. Johns Hopkins, another possible defendant, was seemingly bulletproof: Since nasal radium was accepted treatment at the time, proving private negligence would be nearly impossible. And because the CDC had dismissed the issue, attorneys had little confidence that they could prove long-term damage. Farber throws up his hands. "They all said it was a legal morass," he says.

He had high hopes for a doctoral thesis published by a Johns Hopkins researcher in 1997. In it, Jessica H.C. Yeh surveyed patients treated with NRI by Johns Hopkins physicians at a Maryland clinic between 1940 and 1960. Her findings linked cancer and NRI even more dramatically than the Sandler study.

The study was reported in the *American Journal of Epidemiology* in 2001. But the abstract highlights the low rate of reproductive cancers rather than its central discovery: the significant increased risk of other cancers as well as benign tumors. Yeh found that the risk of developing brain tumors, both malignant and benign, was thirty times greater for those exposed to nasal radiation. The risk of salivary gland tumors was fourteen times greater. Irradiated women had a lower, but not statistically significant, incidence of breast and female genital cancers, suggesting damage to the pituitary gland resulting in a reduced level of hormones and a delayed menarche. In addition, thyroid cancer was also four times more common among those who had had the treatments than those who had not, but according to Yeh, that figure is not statistically significant. The dissertation studied 2,925 patients; 904 had received the nasal radium; the remaining 2,021 had not. "This study alone cannot provide conclusive evidence of the causal relationship between [nasal] radiation and cancers," wrote Yeh. "Nevertheless, along with similar observations from other studies, such a conclusion is reasonable."

To Farber, the paper was a smoking gun. Extrapolated to the entire treated population, he says, over 10,000 excess brain cancers or brain tumors are likely among 1 million treated children.

Yeh does not agree. Because the treatments were variable, and were given at different intervals and at different lengths of time, she says, one can't assess the risk for those who got the treatment outside the clinic she surveyed. "But I agree that these studies suggest that something is probably going on with the treatment," she adds.

How widespread is the problem? That remains to be determined. Most distressing is the failure to be more careful regarding exposure of young children to the ill-understood effects of radiation. As Dan Guttman, the former executive director of ACHRE, asks, "Were these researchers sufficiently cautious as they began their work with children? It's a tough call."

By today's standards of informed medical consent, nasal radium irradiation would be regarded as a flagrant abuse of trust.

Not surprisingly, the CDC has preferred to sweep doubts about the procedure away, confining its warnings and assistance to the squeaky wheel of veterans' groups. And the government hasn't seemed to help matters with its unusual steps—acknowledging risks of the treatment for veterans, underrating them for civilians.

But much remains the same. It took the Food and Drug Administration until 1998 to require drug manufacturers to provide information about how their drugs can safely and effectively be used on children. Testing drugs on children is more complicated than on adults; they often metabolize or absorb drugs at a different rate, and suitable doses are therefore difficult to estimate. Still, they are prescribed medications for a host of illnesses—including nasal disease. Children over four are routinely prescribed antihistamines and nasal steroids.

Farber, of course, is deeply cynical about this chapter of dubious medicine, as well as the climate in which it took place—and now continues. If the medical establishment could successfully "spin" past treatments, he wonders, what prevents them from doing the same with those being offered now? "Today's research institutions are no different from the government during the Cold War," he says ominously. Surely, the back-and-forth about treatments for conditions from back pain (surgery is the answer; surgery is not the answer) to cancer (early mammograms save lives; early mammograms are a waste of time) do give one reason to lack confidence, even in an era of increased accountability and openness. "Nothing has changed," he says.

But from the nose's perch, at least, something did change. By the end of the twentieth century, the idea of the nose as a problem to be solved was joined by the notion of exploiting it—as a resource, as a key to understanding the brain, and as a commercial tool.

Part Two

The Nose and Modern Science

five

What Smells?

> ". . . the nose acts always as a sentinel and cries, 'Who goes there?' "
>
> —Brillat-Savarin

By the late twentieth century, researchers had begun to have some clearer idea of how the nose worked, and how everything from viruses to vicious odors could attack it. The mechanism was one that could scarcely have been appreciated by early pioneers in the field. In order to understand how we smell, scientists first had to understand the molecular workings of the brain itself.

A cramped lab at Columbia University's medical school is home to many of the most crucial findings. In an office high above Manhattan's Washington Heights, tiny white mice—tomorrow's experiments—squeal in cages, and young doctoral students rush past each other in the halls with frenetic urgency.

It is a muggy day in late summer, and from the sliver of the window, you can see the steel magnificence of the George Washington Bridge. Dr. Richard Axel, a tall, terse molecular biologist and biophysicist who directs the lab, chain-chews Nicorette as he glides purposefully through the corridor.

A decade ago, Axel and his colleague Linda Buck discovered that as much as 3 percent of the human genome—an extraordinary figure—is devoted to identifying odors. Humans can detect up to 10,000 different odors, which we achieve by having 1,000 different protein receptors—each of which requires a gene to encode it. (There are an estimated 30,000–100,000 genes in the human genome.) Recent estimates show that humans have roughly 5 million olfactory receptor cells, about as many as a mouse. A rat has some 10 million, a rabbit 20 million, and a bloodhound, which can track scents for longer distances than most mammals, has as many as 220 million.

Many human olfactory genes are more than 10 million years old. Some scientists suggest that not all of them function today (one estimate is 400). Our prehistoric ancestors relied on odors to live, but as the intense need for smelling faded over time, some of the genes mutated. Today many of them contain major defects, but are still recognized as olfactory genes.

Axel wanders throughout the lab with a menacing scowl, checking his breast pocket for his blister pack of gum. After a brief meeting—he makes clear he has little patience for writers—he asks his graduate students to show me something "little" and "pretty." Joseph Gogos, a thirty-eight-year-old geneticist and neurologist from Athens, looks up from his lens, and motions me to approach him. He has thick dark hair that is threaded with gray. "Do not pay attention to him," he says with a shy smile. "No one is smart enough for him." (Such a declaration is hardly reassuring: Gogos earned his M.D. from one of the world's oldest medical schools, at the National University of Greece in Athens, and got his Ph.D. from Harvard.)

When he talks about the olfactory bulb of mice, Gogos speaks so quickly he drops Greek words into his sentences. "Look!" he shouts, instructing me to peer into the scope. "Isn't it just beautiful?" Before me, on a slide smeared with formaldehyde, is a slice of a mouse's olfactory bulb—a dirty white triangle spattered with jagged blue lines. Dozens of the turquoise scrawls end in a single point, resembling lightning that has been captured on film.

The lines are neurons leading from the animal's nose to its olfactory bulb—a "map" discovered by Axel and Buck that highlights the first step of the olfactory process. By tracing the path of these neurons, they discovered that those responsible for a single type of odor molecule reached a single point in the olfactory bulb. "Isn't it amazing?" Gogos asks quietly.

On a yellow legal pad, Gogos depicts what happens when an odor molecule hurtles through the air, the nose, and finally the brain. He draws concentric circles: the nose, the olfactory cortex, the brain, and the human head. Beneath them, he sketches a square Chanel bottle.

When we smell perfume, odor molecules from the liquid bind to receptors on a dime-sized patch of tissue at the very top of the nose called the olfactory epithelium. The receptors are part of neurons that extend three to four centimeters from the epithelium to the brain. Unlike other neurons, which are encased in the skull, those in the olfactory epithelium are exposed to the air we inhale. While there is significant scientific debate about whether or not brain cells regenerate, researchers agree that olfactory neurons replace themselves every two months or so. A layer of stem cells beneath them creates new olfactory neurons, maintaining a healthy supply.

Each olfactory neuron in the nose has a long fiber, or axon, that pushes through a tiny space in the bone above it, the cribriform plate. There, it makes a connection with other neurons in the olfactory bulbs, two cylinders about the width of a pencil that rest behind the gap in the eyebrows. Like a telephone switching point, the olfactory bulbs are the site of key connections; from there, impulses are relayed to the brain's limbic system, which governs emotions, sexuality, and drive, and the hippocampus, which is thought to encode the information into memory. Connections between the olfactory bulb and the neocortex, the part of the brain responsible for thoughts, language, and behaviors, are thought to be more circuitous.

The neurons are placed randomly throughout the epithelium. Yet when the axons reach the olfactory bulbs, those that picked up a signal from the same receptor converge on the same place in the olfactory bulb—the tiny blue point I saw in the slide.

Most odors consist of mixtures of different molecules, meaning that the brain "reads" different odors by the receptors it sets off. "The neurons tell the brain, 'I see something at point A, point G, point X,' " Gogos says. "Then the brain does the calculation—'if it's A, G, and X, it must be garlic.' This means that the chaos we have in everyday life—all the smells around us—are actually regimented and organized in the brain." We know the pattern of the taste buds lining the tongue—salty and sweet in front, sour on the sides, bitter in the back. Odors reach the brain with a similar structure.

A few months later, we trudged through dirty piles of snow to an overheated café. When the door opened onto the cold street, a rush of odors hit us: ginger, cinnamon, tea, coffee—a perfect mix for Gogos to discuss. "We always use this imagery to describe it, but it seems to work: olfaction is like a lock and a key. The receptor is the lock, and it can be activated by a specific key whose shape fits into the lock's. That lock is a protein with a specific shape and distribution of charged molecules. So different protein receptors can only be matched to different odors. When an odor fits into the right receptor, that protein in turn processes the cell—and that sends messages to the brain that it has 'smelled.'

"A single scent is made up of more than one type of molecule—sometimes even dozens. You might imagine that for every odor molecule, there may be a different receptor, determined by specific genes. But this would be impossible—the brain would do nothing but detect odors! The brain only remembers ones it needed in order to survive, to evolve. That means we keep track of things like ripening food, poisons, mates we respond to."

Each olfactory receptor has the ability within it to bind with several odor molecules. Conversely, each odor molecule has the ability to bind with a range of potential receptors. The intensity of the binding, Gogos says, varies, depending on the quality of the fit. (In fact, each of your nostrils also detects odors differently. This may in part answer why we have two nostrils in the first place—to add to the brain's stereo perception of scent.)

Outside, a mutt sniffs the steel pole of a parking meter, and

Gogos waves toward it. "Most animals are more sensitive to smells than humans, but the brains of all mammals work exactly the same way," he says. Like the blue-tinted signals that fired to the mouse's brain, the dog's brain is being "unlocked" by odor molecules left behind by another dog. "We may think that animals are more in tune to odors, but humans are pretty sensitive to them too. We just stand on two feet, away from most odors. Odor molecules are heavy, and the farther up you go, the fewer there are. The most interesting odors float just above ground. You can see this with dogs and cats—they sniff the ground when they're trying to catch a scent. Our ancestors probably did that, too. Smell itself is a very elegant and orderly process, but somehow the ideas and images that surround it are exactly the opposite."

Darwin, in fact, theorized that the relatively large visual cortex in primates had superseded the olfactory cortex, which had withered over time. He and other scholars believed that our upright posture, which allowed for better visualization of our surroundings, diminished the necessity of detecting odors. Likewise, Freud suggested that our bipedalism was the triumph of sight over smell. And as recently as 1981, the Norwegian neuroscientist Alf Brodal wrote ". . . the sense of smell is of relatively minor importance in the normal life of civilized man."[1] The concept that smell is important, even integral, to our knowledge of the brain—of life—is a new development.

Until the 1980s, scientists say, research involving smell was viewed by many as marginal, even frivolous. "You cannot imagine the reactions you'd get twenty years ago when you'd tell people you were a grown man who studied the sense of smell," says William Cain, an environmental psychologist who specializes in olfaction at the University of California at San Diego School of Medicine. "People just looked at you in disbelief," he says. "You'd have to explain it, spell it out. 'Olfaction. It's the science of smelling.' " Yet today, the number of olfactory scientists who work in academic research, the fragrance industry, or both (often, large companies such as Proctor and Gamble or Nabisco fund research) has risen dramatically from the ranks of just a decade ago.

For example, Monell Chemical Senses Center, founded at the University of Pennsylvania in Philadelphia in 1968 as an institute for multidisciplinary research on taste, smell, and chemosensory irritation, has trained more than 370 scientists. The researchers work in government, industry, and academia in areas as diverse as waste management and fragrance development.

Since 1980, the National Institutes of Health has more than quadrupled its funding for research in olfaction, which went from $4.3 million in 1980 to $25.7 million in 1999.[2] Meanwhile, membership in the Association for Chemoreception Sciences, an organization of researchers from neurologists to zoologists who are involved with taste and smell, has risen from some three hundred a decade ago, to 650 in 2001.[3]

The progression of how the brain turns odor molecules into smells seems straightforward, but research efforts aside, Gogos and other scientists caution that the picture is—as yet, anyway—incomplete. "We don't know all the answers," Gogos shrugs. "It's not like Greek runners dashing through the brain with an odor molecule as their baton," he says.

The brain's pathways are still only partially deciphered; no one is sure how one part communicates with others. But by studying how olfactory neurons make connections with other neurons in the brain, researchers hope they can identify how nerve fibers connect elsewhere, and may point to ways of making neurons regrow in other parts of the nervous system. Researchers are also studying how receptors respond to different odors, which may lead to treatments for people in whom age, disease, or other factors have destroyed or damaged the sense of smell.

While researchers understand some stages of how the brain perceives odors, others remain mysteries. For example, what happens to information about smells after it has made its way from the olfactory bulb to the olfactory cortex? How does the brain process that information? And how does it reach the higher brain centers, in which information about smells is linked to behavior? Researchers in Miami found that adults who sniffed lavender before and after tackling simple math problems worked faster, felt more relaxed, and made fewer mistakes than those exposed to other odors. And in one small British study, elderly insomniacs who sniffed lavender before going to bed fell asleep sooner—and stayed asleep longer—than those using sedatives.

But because the human brain is so complex, many researchers look to animals for answers. John Kauer, a neuroscientist at Tufts Medical School and New England Medical Center in Boston uses the salamander's nasal cavity—a flattened sac beneath the skull—to analyze the entire olfactory system, from the first olfactory neurons to how odors affect behavior. The sac can be opened easily, providing access to olfactory neurons. Kauer devised an optical recording system in order to make observations with dyes and video cameras instead of probes. With it, Kauer says, you can actually watch the brain as olfactory perception takes place.

Research on smell is, of course, far from simply research on smell. As it turns out, the receptors used to detect odors are closely related to the protein receptors in our brains that recognize neurotransmitters such as serotonin and dopamine—those responsible for our emotional well-being. Genes much like those for odor receptors may, it is thought, control other types of chemical sensing, such as the ability of sperm to locate an egg. Similar receptors also may function in a special structure in the nose called the vomeronasal organ. They help to detect special chemical signals called pheromones, which regulate hormone release, mating, and social functions in animals—and possibly in humans. "Learning about how odors affect us isn't simply learning about how odors affect us," says William Cain. "It's learning about how we live."

Scientists say that one of the most complicated and challenging aspects of research on smell is assessing how, in humans, the sense interacts with reasoning and emotion—and how those, in turn, affect smell.

Our ability to detect scent is the most ancient sense. Even the most primitive organisms had an olfactory system that detected food, poison, and predators. Imagine the brain as an archeological site, with three layers. The oldest part, the brain stem, is sometimes called the "reptilian" layer, the remains of our days as amphibians. It controls basic functions like heart rate, digestion, and breathing, and it is also the main channel for sensory and motor signals. Built like a thick stalk, the brain stem originates in the spinal cord and reaches up to the next layer, the limbic system, which encases the amygdala, two almond-sized organs largely responsible for feelings and temperament. In a bow to its close connection to smell, the limbic system was originally called the "rhinencephalon," literally, the "nose-brain." The neocortex, the newest part of the brain, is the outer two-thirds of the organ; language and reasoning are processed here. The early brain's ability to decode smell, in fact, helped spawn our ability to feel and rationalize. As the poet Diane Ackerman wrote, "We think because we smelled."

But smelling something, or thinking about it, doesn't mean we can describe it. In fact, most people can scarcely even characterize what they smell. When sniffing, say, a jar of peanut butter, sweaters stored in a mothball-filled trunk, or a charcoal grill on a hot July afternoon, people are apt to turn the noun they're smelling into an adjective. Peanut butter smells . . . nutty. The sweaters? Stuffy—mothbally. The grill, for my husband, summoned the following depiction: "Acrid? Sweet? Hot?" Ask a person to describe his or her worst blind date, or even fish they saw while scuba diving, and get an animated recall: a stubbly goatee, neon green scales. With smells, however, even the most articulate among us become tongue-tied.

For years, scientists attributed this to the lack of "wiring" between the part of the brain that recognizes odor and that which processes language. Another common belief was that smell had a greater influence on emotions than other senses because of the olfactory cortex's proximity to the limbic system. New technologies that allow scientists to glimpse the brain at work have challenged those views, which date to the 1930s and 1940s. Using equipment such as functional magnetic resonance imaging (fMRI), which provides a continuous picture of brain activity, and evoked potentials, a test which charts the electrical activity of the brain with electrodes, researchers have documented interactions between the olfactory cortex and other parts of the brain.

Tyler Lorig, a psychologist at Washington and Lee University in Lexington, Virginia, has been searching for the external manifestations of those connections for twenty years. A soft-spoken man with a salt-and-pepper beard and a gentle Georgia drawl, Lorig spends his days in his lab, "thinking about smelling." He believes that one reason we are at such a loss to describe smells is because the brain's mechanisms for odor compete, quite literally, for brain power—especially that used for language. "They actually interfere with each other when they are simultaneously processed," Lorig says.

"Picture the brain like an orchestra, with a trumpet section, violins, clarinets—the works. When you tell the brain to both simultaneously smell and describe that smell, it's like telling the violins to play two different melodies at once," he says. "You can get by, but the performance is degraded."

Lorig believes the phenomenon is rooted in part in industrialized Western society simply not "developing" its sense of smell. Even the English verb "to smell" is imprecise. It is both transitive and intransitive; a person can simultaneously smell bad and smell something bad. Other cultures, and thus their languages, treat scent differently. Some traditional societies, such as the Kapsiki of Nigeria and Cameroon, have specific vocabularies to describe odors, from "old grain" to "white millet beer." Likewise, the Desana of Colombia have odor hierarchies for "bland" and "strong" smells.[4]

Yet many Westerners struggle to describe the nuances of odor. The notion that culture holds our olfactory capabilities back is clear in the case of Stephen D., a young medical student the neurologist Oliver Sacks describes in *The Man Who Mistook His Wife for a Hat*. One night, Stephen dreamt he was a dog. The dream was so redolent with aromas that upon waking Stephen actually smelled with a perception he had never experienced. He found himself sniffing perfumes in fragrance shops, and recognizing friends and patients by their scents. Even days afterward, he was finding his way around Manhattan by odors, like a German shepherd.

Sacks attributes the heightened awareness—disinhibition, he calls it—to the drugs Stephen regularly took, likely amphetamines. And indeed, three weeks after his unusual ability to smell appeared, it waned just as startlingly. "That smell-world, that world of redolence," Stephen exclaims. "So vivid, so real! It was . . . a world of pure perception, rich, alive, self-sufficient, and full. If only I could go back sometimes and be a dog again!"[5]

In a society dedicated to the suppression and eradication of odors, it is little wonder that we lack a rich lexicon for what we try to abolish. Experts in the fragrance industry—known straightforwardly as noses—can distinguish minute changes in thousands of odors in devising new perfumes. Yet such sensibilities are far from the norm. While most humans are capable of identifying and labeling odors—the perfume industry alone channels $6 billion through the U.S. economy—they are rarely encouraged to exercise their sense of smell.

Lorig points to himself as a perfect example. "For twenty years I've been around smell labs and ideas and theories about odors. I've tested my sense of smell dozens of times, and the truth is, I'm average.[6] But because I'm accustomed to smells, and identifying them, I've gotten pretty good at being able to label them. It's all part of a larger question about smell. There are people who can perceive odors with more precision than others, sure. But the truth is, noses at perfume houses and vineyards are simply above average—they're not off-the-charts in terms of their abilities. What's different about them is that they *use* their abilities better. Was Michael Jordan always destined to be the greatest basketball player? Well, that depends. He had the raw talent, but he also put it to use. It's the same thing with people in this line of work—if you're going to come up with new scents, you have to have a good nose. You can't have an anosmic putting together perfumes. But it's also what you do with what nature gives you, how you develop it."

A glance at our "olfactory education"—this is what researchers like Cain and Lorig call it—shows that the chances of developing that are slim. Toddlers don't sit in front of games that waft odors—they learn to distinguish triangles from squares, moos from barks. Kindergartners don't differentiate between rose petals and lily of the valley; they learn the alphabet and musical notes. "When a kid goes around sniffing something, the parent doesn't say, 'Oh, good boy, Johnny, you're picking up on odors!' " Lorig says. "The parent says things like, 'That's not polite.' Somehow, we get the message early on that our sense of smell is primal—too primal for polite society." The result, Lorig says, is a fundamental portion of the brain that science—and possibly even learning—have overlooked.

Lorig's research underscores how humans repress their awareness of odors. In one recent study, he and his colleagues told ninety-three subjects that the experiment in which they were about to participate was designed to "evaluate the effects of sensory conditions on their judgments about art." Yet only three subjects noticed that odor was involved in the experiment—despite the strong smell of lavender or vanilla emanating from behind the slide projector. One participant in the vanilla room said she thought lighting levels were being manipulated—then told researchers she felt like going back to her dorm "to bake some cookies."

"When I heard her say that, a big old smile just crossed my face," Lorig recalls. "I just knew she had to be in the vanilla room. But the real question is this: why does the brain play down this information? What's going on?" One possible explanation could be the space and time at which the brain "received" the message. Perhaps the student was concentrating on the colors on the screen, and the odor molecules from the vanilla reached her just as she was trying to decide what to say about the artwork. "It could be simply that the brain experiences a temporal overlap—it's trying to do two things at once, and can't."

In another experiment, Lorig and his researchers compared the brain's electrical responses to odors and tones. All stimuli, whether delivered through sight, smell, taste, hearing, or touch, evoke minute electrical signals, which travel along nerves and through the spinal cord to specific regions of the brain; electrodes attached to the scalp register the data.

The researchers asked the subjects' ability to judge the size of objects, words, and numbers that flashed on a screen as odors and then tones were piped into the lab. Electrical impulses were collected as the subjects calculated whether the item on-screen was larger than its predecessor.

The researchers found that when the tone was played, the brain's electrical activity was similar, whether the image was displaying a thing, a word, or a number. But when odor was introduced, pictures had a different effect than words or numbers. "We don't know why, but the brain would then go about making the decision in a different way—activity would shift. It didn't disrupt the ability to finish the task, but it did alter the impulse that went together to complete it."

Does this study explain how some people can "block" out foul odors? On the commuter train that takes people from my suburb to Manhattan, I am always amazed that anyone can sit in the same car as the grungy bathroom, let alone right outside it. Yet people plunk right down without seeming to notice. Once I had a dermatologist's appointment in a building a few floors up from a Chinese restaurant. I walked in, sat down, picked up a magazine, and was so repulsed by the odor of old fried shrimp that I just couldn't bear to be there (although arguably I was not eager to have a wart burned off my foot). The doctor was forty-five minutes late, so eventually I feigned indignation about the fact that civilians have schedules, too, and left. But the truth is that I was too embarrassed to say that the place reeked—it seemed unduly insulting to the poor receptionist. Yet the waiting room was full.

Studies indicate that, depending on a person's view of a particular odor, he or she may become temporarily desensitized to it. In other words, if you are a hog farmer who makes a living selling pork meat, or work for a nice dermatologist who pays you generously but who happens to be upstairs from a smelly take-out place, chances are that, in time, you will become habituated to offending odors. "Your 'baggage' about a certain scent will influence how your brain reacts to it," says Charles Wysocki, an anatomist and psychobiologist at the Monell Chemical Senses Center in Philadelphia. If you're a trash collector, over time the brain may well "learn" that the scent of rotting garbage is not harmful. (The same effect holds true for perfume counter employees.) Since one of olfaction's biggest functions is to sniff out danger, the brain may simply set up "filters," Wysocki says, that let that malodorous information pass right through, unnoticed. If, on the other hand, you live downwind of the hog farm, and are convinced that it's the cause of your headaches, you may find it impossible to ignore.

But how does this explain the people who sit next to the bathroom? Is it poor acuity? Or a good ability to block out negative sensory information?

"That's one of the great unanswered questions about smell," Lorig says. "We just don't know how odors—good or bad—affect concentration." When odors are introduced while subjects are reading, they are barely noticed. It is as if their channel to the cortex is blocked by attention to reading. This is one of the reasons why I think constant exposure to a smell doesn't interrupt cognitive processes. In our experiments, we irregularly presented very short bursts of odors. We didn't give people a chance to get used to the odor so each time the odor came in, they had to deal with it. That's why I think we saw a change in the way the brain processed the nouns. If we had done the experiment with constant odors, I don't think we'd see a difference."

Lorig acknowledges that scientists' understanding of these issues is in its infancy, but is confident the brain will someday yield its secrets. When it does, he says, the current state of understanding will be seen as cartoonish. "We'll look at this and think 'Man, we were using leeches,' " Lorig says. "But that's the only way to get from here to there."

There is almost certainly a connection between how mankind evolved and the effect smells have on brain functioning, but it remains unclear what was cause, and what was effect. "Since so much of human behavior, and arguably all cognition, is filtered through language, our limited language for odors may be a cause of our disregard of this sense rather than an effect of our getting our noses off the ground," Lorig says. "Maybe it even worked the other way: standing on two feet reduced the olfactory information load and allowed language to flourish. Either way, odor and language don't seem to work together."

One clue to the centrality of brain activity and olfaction is that aging and many neurodegenerative disorders have a measurable impact on the sense of smell. "We're not sure why smell deteriorates with age, but pictures of the brains of older people show atrophy of the olfactory bulb and some reduction of the receptors," says William Cain.

As researchers study smell and aging, several truths have emerged, Cain says. Age takes a much greater toll on smell than on taste. Women at all ages are generally more accurate than men in identifying odors. Smoking can adversely affect the ability of both men and women to identify odors. Although certain medications can cause smell and taste problems, others—notably antihistamines—seem to improve the senses of smell and taste, he says.

Viral infections, sinusitis, and certain medications can impair your sense of smell, as well as tobacco smoke and other irritants. Workers exposed to harsh chemicals, such as those in paper mills and chemical manufacturing industries, often have compromised olfactory ability. Hunting dog pens are warned not to clean the close quarters of dog pens with harsh cleaning products, for fear of damaging the dog's ability to pick up scent. Dairy and livestock farmers also risk numbing their sense of smell: constant exposure to the urine and gases from manure can diminish sensitivity.

People with certain neurological disorders, such as Parkinson's and Alzheimer's, also suffer smell loss. Of all the indignities of Alzheimer's—and they are legion—perhaps most striking is the frightened confusion of those stricken with the disease. Words become impossible to summon. Sounds are threatening. Faces of friends and relatives, some beloved for a half-century, are as indistinct as blurry crowd photographs. Odors, too, are lost.

In a study at Columbia University, researchers administered a standard smell identification test to seventy-seven people with mild cognitive impairment (MCI) whose symptoms also included moderate memory loss. Studies have shown that people with MCI develop Alzheimer's at a higher-than-average rate, and researchers hoped that the test might be a valuable early detection test. Test subjects were given cards scented with familiar odors such as menthol and peanuts, and were scored on how well they could identify those odors. Two years after taking the test, those who had scored lowest were much more likely to develop symptoms of the disease than those who scored higher.

Proust wrote that when nothing else exists from the past, after people are dead and objects are broken or lost, "the smell and taste of things remain poised a long time, like souls bearing resiliently, on tiny and almost impalpable drops of their essence, the immense edifice of memory." Now, too, it turns out, Alzheimer's strips us even of the core human perception of smell.

The Nose and Sex

Because modern science of the nose is so new, there is, as one might expect, a great deal of controversy among researchers. While the work of Axel and Buck is universally acknowledged as fact, many who work to decipher the inner workings of the nose have received far less recognition for their labors. Some, for instance, work to solve mysteries such as anosmia, the inability to smell, but are seen as downright quacks. Yet no few millimeters of nasal tissue stirs controversy quite like the vomeronasal organ, or VNO, which some say detects pheromones.

David Berliner, an anatomist, tends to be controversial. As a young professor at the University of Utah in the 1960s, Berliner worked with substances he had found in human skin. In those days, skin cells were plentiful in Salt Lake City, surrounded as it is by ski resorts—and grounded skiers with broken limbs. Dead skin accumulated in plaster casts in such abundance that, at least to Berliner, it seemed a natural thing to study. So he leached some of the cells and put them in solvents. To his surprise, when he left some of the extracts in open vials around the lab, he noticed a sudden—and puzzling—rise in camaraderie among a group of coworkers. Normally, they snapped at each other, and ate their sandwiches hunched over papers in their cubicles. Yet when the liquid skin extracts were left out in the open, Berliner's colleagues suddenly laughed, joked, ate lunch together, and even played bridge. Strange, Berliner thought, as he observed the group. Some months later, he put the containers away, and the group became churlish again.

Researchers had just begun to identify pheromones (the word is derived from the Greek *pherein,* "to transfer" and *horman,* "excitement") in creatures ranging from ants to elephants. Berliner theorized that the extracts contained pheromones, chemicals given off by insects and mammals that affect only others of the same species. In many species they reach the brain by way of the VNO, which transmits signals to the brain.

Berliner became convinced of the importance of pheromones in human behavior. Other research on the topic was proving intriguing. In the 1970s a Wellesley undergraduate named Martha McClintock observed that many women in her college dormitory had synchronized menstrual periods, and documented the phenomenon in her senior thesis. At the same time, Winnifred Cutler, a biologist and behavioral endocrinologist, was finding that women with active sex lives have more regular menstrual cycles than those with only sporadic sexual encounters. Both researchers believed pheromones played a role in their findings.

The VNO and pheromones had been discussed in scientific literature since the early eighteenth century, when a Dutch doctor named Frederick Ruysch identified the human VNO, a bilateral pea-sized cavity lodged in the nasal septum. In modern times, though, scientists assumed that it had become useless, like the appendix, with no apparent function. Chemical communication in the animal world is complex and varied; many mammals can detect pheromones and other signals through their glands, urine, and saliva, as well as the VNO, which is connected to parts of the brain involved in reactions rather than cognition. But in humans, the links from the VNO to the brain, and behavior, are much less clear.

For decades after her groundbreaking thesis, McClintock, now a psychologist at the University of Chicago, devoted her research to identifying the mechanism behind the timing of the menstrual cycles. She and others believed that pheromones were the cause, but evidence was elusive. Berliner, who shortly after his stint in Utah left research behind for commerce, had put his curiosity about the substances on hold. As a Silicon Valley venture capitalist in the early 1990s, Berliner still dreamed of bridging the gap between business and academe, and the possible role of (and market for) human pheromones. By then, the human nose was attracting attention both in science and the marketplace. Fragrance, and its emphasis on sexuality, was ubiquitous. So was talk of animal pheromones. In the mid-1980s, researchers using an electron microscope discovered the existence of the human VNO in more than two hundred adult subjects. New findings on animal pheromones were being published all the time. One study showed that a male pig breathes a chemical called androstenol into a sow's face; she then spreads her legs so the male can mount her. Queen bees use pheromones to thwart the sexual maturity of female worker bees, and animals from crabs to marmosets release pheromones to mark their territory.

Finally, Berliner thought, the time was right for a look into the extracts he had saved and frozen from the Utah lab—the memory of experience had left a lasting impression. "The change in attitudes and emotions of my colleagues was extraordinary, it was impossible to forget. I kept coming back to it," he says in his Spanish-accented English. The son of Eastern European Jews who emigrated to Mexico, Berliner had a unique childhood. He was surrounded by independent thinkers, writers, and artists; Diego Rivera and Frida Kahlo were often guests in his parents' home. A signed Rivera print—"To David"—hangs prominently in the entranceway of Berliner's Menlo Park office. Huge pink oleander blossoms wave outside his window.

The more scientific papers he read, the more Berliner was convinced that humans responded to pheromones as well. Was it possible that the skin of certain people emitted a substance that could alter the moods of others, he wondered? Berliner reasoned that the centimeter-long VNO acted as a sensor for airborne human pheromones. As in animals, he thought, the VNO provided the pheromone molecules a direct line to the hypothalamus. Though tiny, the size of an unshelled peanut, the hypothalamus has many functions: it regulates body temperature, blood pressure, heartbeat, appetite, thirst, the metabolism of fats and carbohydrates, and sugar levels in the blood. As part of the limbic system, the hypothalamus is also thought to be the principal site for emotions such as fear, joy, and anger, as well as sexual behaviors. Structurally, it is joined to the thalamus; the two work together to monitor the sleep-wake cycle.

But while the VNO is clearly present in the human fetus, it has significantly atrophied in size even by the time a baby reaches full term. Many scholars questioned its function. Humans still have wisdom teeth, but that doesn't mean they need them. Still, Berliner was determined, and enlisted colleagues back in Utah to investigate his skin extracts, among them Luis Monti-Bloch, a Uruguayan neuroscientist at the university.

Monti-Bloch, an elegant, soft-spoken man in his early fifties, measured electrical activity in VNO tissue in response to chemical stimulation. While fragrant compounds produced no electrical effect from the VNO, Berliner's skin extract, on the other hand, made it activate. Berliner and Monti-Bloch isolated a substance from the skin of males they called androstadienone. When they applied the substance to the VNOs of women, the women showed signs of being more relaxed: they experienced a decrease in their heart and breathing rates, and their body temperatures rose. The team was thrilled. This, Berliner says, likely explained the experience in the Utah lab. (A similar substance, estratetraenol, extracted from the skin of women, was also identified.)

I visited Monti-Bloch at a Utah lab on a cold February day in 2000. The lab, sequestered in a Salt Lake City office park at the foot of the Wasatch Mountains, seems an unlikely place for unlocking the mysteries of sex and desire. The conservative center of the Mormon universe is better known for its choir and caffeine-free living than for its exploration of human appetites.

"The VNO is truly the body's sixth sense," Monti-Bloch says, pointing to an enlarged picture of the organ on his desk. "It's right here," he says excitedly, "right here, in almost everyone." The organ can be inadvertently removed during nasal operations. I mention that I've had four sinus surgeries. "I hope its still there," he says. "Let's take a look." He gestures to an examining table, and I hop up onto it. A graduate student from Mexico looks on, and Monti-Bloch peers into my nose with a tiny scope. "It's there!" he exclaims. "You have it!"

What a relief. But I knew I hadn't lost my sixth sense. Once, on a near-empty train from Baltimore to New York, I moved away from a man, who, an hour later, raped a woman in the bathroom. Another time, I stood on the landing of my apartment, which looked absolutely normal. But I somehow knew that I had been burgled. Something told me to put down my keys and find a neighbor before going in. I tell Monti-Bloch of these episodes. He informs me that my VNO was picking up on pheromones, and that I reacted. Pheromone researchers are deluged with tales like mine. One Pennsylvania woman, who had tried for years in vain to have a child, experienced pregnancy symptoms—and even tested positive on pregnancy tests—when she had traveled, and roomed with, pregnant colleagues. Both times, she says, the women were so early in their pregnancies they didn't yet know of their condition. Her doctors were at a loss to explain it. Monti-Bloch, on the other hand, thinks her VNO was picking up on pregnancy pheromones released through the other women's skin. "I cannot explain more," he says. "Just that it happened—and that there is so much more to learn."

Berliner believes that one reason mainstream scientists are skeptical of his work is because he became a millionaire in business, seeking profit from commerce rather than government grantsmanship. But now, he stands poised to get his due. Proof of human pheromones received a big boost when McClintock, the Chicago psychologist, published a 1998 paper showing definitively that women respond to the pheromones of other women. She took sweat samples from underarm pads worn by young women who had not yet ovulated that month, and wiped them on the upper lips of study subjects. The pads accelerated the surge of LH, or luteinizing hormone, which triggers ovulation, cutting the subjects' cycles by as much as two weeks. McClintock also found that subjects who were exposed to women who had already ovulated had their cycles delayed by as many as twelve days. (In men, LH stimulates testosterone. Researchers say that pheromones may explain why men find women who are mid-cycle—and most fertile—more attractive than during other times of the month.)

In 2000, neurogeneticists at Yale and Rockefeller University announced that they had found the first human gene associated with the function of pheromones. Researchers isolated a human gene, labeled V1RL1, they believe makes a pheromone receptor. Pheromones, they say, attach to this receptor when they are inhaled into the mucous lining in the nose.[7]

While many researchers say the report promises to settle the pheromone existence question once and for all, others still wonder. Catherine Dulac, a professor of molecular and cellular biology at Harvard, is also dubious. The brain might be subtly informed about the presence of pheromones, she says, but humans are so influenced by other stimuli—and experience—that the leap is just too big.

Meanwhile, Berliner continues to mine possible uses for synthetic pheromones, which he calls vomeropherins. He believes that the studies about pheromones and menstrual cycles suggest that vomeropherins might one day be used as fertility treatments for couples who want to conceive, or as contraceptives for those who don't. Couples who are having sexual problems might be able to use pheromones combined with traditional therapy to enhance desire. The Utah team, which also pours its efforts into developing these drugs for a company Berliner owns called Pherin Pharmaceuticals, are further convinced that vomeropherins can enhance mood, alleviating depression and anxiety. Berliner even thinks that the drugs could one day be used as a preventative for prostate cancer.

Large pharmaceutical companies are encouraged enough by the Utah lab's findings to support development of vomeropherins. Johnson & Johnson recently agreed to fund trials for two antianxiety agents—one for women, one for men. Berliner says the antianxiety agent works as a simple prophylactic. "If you have to give a speech, and are terrified of facing crowds, you sniff the drug and immediately feel better," he says. "The panic dissipates immediately." The synthetic pheromone is delivered in a nasal spray, sending signals to the hypothalamus to relax. Dosages are minute—mere picograms (one-billionth of a milligram)—and, because they are not ingested, have no toxic effects, Berliner says.

Another pheromone-based drug is aimed at helping women with PMS. That drug, funded by Organon, a Dutch pharmaceutical company that specializes in reproductive and psychiatric medications, is in Phase II trials. (The FDA has three categories for human drug trials, each used on wider numbers of people.) Once a drug succeeds in entering the marketplace, there's no telling how it will do financially. But the antidepressant Paxil, which is used by people who suffer from the same social phobias Berliner hopes vomeropherins can treat, had revenues of $1.9 billion in 2000.

A study performed by Martha McClintock and Suma Jacob added more weight—at least for women—to findings on androstadienone and estratetraenol, the substances produced by men and women, respectively. (In the paper, the researchers referred to the substances as steroids.)[8] The researchers used a series of psychological tests to assess both physical and mood changes in subjects after their exposure to each steroid. Both elicited an immediate, short-term positive mood in women, but a negative mood in men. Despite the clear sex differences in responses to steroids, the authors concluded that steroids are not able to trigger specific behaviors in humans. "It is premature to call these steroids human pheromones," the authors wrote, adding that they were nonetheless "psychologically potent"—and mandating future research into their actual function. Indeed, Jacob and McClintock went on to show that these compounds have widespread effects on the brain, even when they can't be detected as odors. "Clearly unconscious odors, if not pheromones, play a much bigger role than we thought."

But others see assurance in studies showing that we may

glean clues about the immune systems of potential mates through their odors—or perhaps odors packed with pheromones. (Some researchers believe that human armpits produce both in large quantities. Pheromones mix with sweat, evaporate, and float into the air if they aren't washed away.)

In the late 1990s, Carol Ober, a geneticist and gynecologist at the University of Chicago, concluded that women choose mates with genetic backgrounds different from their own. She theorized that the behavior tended to give their offspring a strong immune system.

Ober conducted a study of South Dakota Hutterites, a small, tight-knit religious community in the American Midwest and Canada who trace their ancestry to a handful of the faith's early followers in sixteenth-century Moravia. She wanted to examine their MHC, or major histocompatibility complex. Each person has a unique combination of MHC genes that encode various components of their immune system. The Hutterites are ideal for MHC study, since they marry only within their small community (the entire group, dispersed throughout the two countries, numbers only 35,000; Ober's research into their genome is ongoing). When they are of marriageable age, single men and women visit nearby Hutterite colonies, and meet potential spouses during holiday celebrations or farmwork. And since they don't use perfumes or deodorants, their odors—and pheromones—are in abundance.

Ober wondered if the Hutterites naturally chose partners with dissimilar MHC, and examined 411 couples drawn from thirty-one Hutterite colonies. While the genetic variation among the group was obviously limited, only forty-four couples, roughly 11 percent, matched for certain types of MHC. The chances of this happening randomly are only about 5 percent. Ober believes that the study's results show that MHC genes may actually influence mate choice.

Ober also found that those who had similar MHC tended to have longer periods between pregnancies and higher rates of miscarriage. Fetuses who receive the same MHC genes from each parent, Ober believes, may be more likely to abort, perhaps because of some unknown immunologic mechanism. The more varied a person's MHC, the more robust its immunity.

In similar research, in the mid-1990s Swiss zoologist Claus Wedekind sought to see if clues to genetic dissimilarity lay in body odor among forty-nine female subjects at the midpoint of their menstrual cycles, when olfaction is at its keenest. Then they were asked to sniff six sweaty T-shirts from male subjects. The women, who had given DNA samples, were asked to "rate" the T-shirts for intensity, sexiness, and pleasantness. Three T-shirts were from men whose MHC were similar to their own, and three from men who were less similar. Women found the odors of men who were genetically dissimilar more pleasant than those who were similar—unless the women were on birth control pills. When the women were on oral contraceptives, which can mimic the first trimester of pregnancy, they preferred the scent of men who were genetically linked to them. Wedekind says the results may indicate a throwback to ancient women, who relied on next-of-kin for protection during pregnancy. Pregnant women—and women whose bodies are tricked into thinking they're pregnant—might have the same tendency, Wedekind says.

Of course, humans make decisions about their mates based on a vast number of cues, from looks to financial security. Do the effects of the Pill really scramble normal signals of mate selection? "It's a reasonable question," says Monti-Bloch. "Whether it is sabotage to the reproductive system, is impossible to say."

But what is it that is actually attractive about odor—or pheromones—and what do they have to do with love and sex? Sweat glands in the armpit area are only active during the times in which the human being is capable of reproduction. They first kick in during puberty, and slow down during menopause. Researchers say it's no accident that during intercourse or even embracing the woman's head (and nose) are frequently near the man's armpit. Many say that this physical proximity is essential for the transmission of odors—and pheromones. (Heavy winter clothing obscures these signals, and the frequent scrubbing Americans generally do may wash them away altogether.)

Monti-Bloch believes another pheromone-rich site is the nasal groove that runs from the outside of the nostrils to the corners of the mouth. Kissing, he believes, is a ritual that developed to detect pheromones.

Whether synthetic pheromones alter the way we react to events and people remains to be seen. The human brain is complex, and, as everyone knows, there is much more to sexual attraction than airborne particles. Humans respond to sights and sounds (not to mention societal pressure and fat bank accounts). If they didn't, movie stars would be unlikely to attract fans, pornography Web sites wouldn't keep the Internet financially afloat, and glossy magazines with pretty models on the cover wouldn't sell.

Even so, evidence continues to mount that odors affect our sexual decision-making. Rachel Herz, an experimental psychologist at Brown University who researches the effects of odors on behavior, asked 166 women what makes a man attractive enough to consider sleeping with. The women said that appearance mattered, of course, as well as the sound of a man's voice and the touch of his skin, but scent was ranked as the most important, particularly in deciding whom *not* to sleep with. It's not as if women are jumping out of their clothes whenever they get a whiff of someone who turns them on, Herz explains. "It's more like, 'You don't smell right, I'm not going ahead with this.' "

The preferences of the human nose, of course, are complicated. Researchers at Monell once found that foul odors provoked happy responses in study subjects just as often as pleasant ones. Some people are fond of the smells of animal manure or even dead skunks, at least in passing. And body odor itself is hardly static: it depends on food and genes—and may vary with fear, happiness, or anger.

Science has yet to determine just how crucial the ability to detect odors is to living: to sex and love, to eating and remembering, to inspiring and to bewitching. Nowhere is this more evident than in Patrick Suskind's novel *Perfume.* The book's hero, Jean-Baptiste Grenouille, is born in the slums of eighteenth-century Paris with an uncannily acute sense of smell. Grenouille lacks a scent of his own, which renders him a pariah from birth onward: his mother, disgusted with his odorless body, abandons him in the gutter at birth. Apprenticed to a master perfumer, Grenouille's gift spawns an obsession to create the greatest fragrance the human nose has ever known. But to do it,

Grenouille must actually obtain these odors, the best of which reside in the bodies of beautiful young virgins. He becomes a serial killer, slaying girls in order to possess their scents. Grenouille, this olfactory vampire, has a barbaric triumph as he terrorizes France in the elegant "Age of Reason." He is finally caught and tried in Grasse, the center of perfumery, but once more his freakish acumen assists him: he dabs himself with his concoction, which proves so seductive that the scene for his execution becomes a frenzied orgy.

Suskind's message—that even in the age of logic and abstract thinking, the nose, this "primitive organ of smell, the basest of senses," as a priest in the book put it, held sway in inexplicable ways. *Perfume,* which topped international best-seller lists for months, shows that even the thought of scent's powers captivates us still.

six

The Recreational Nose: A Glimpse

> Nobody saves America by sniffing cocaine, Jiggling yr knees blankeyed in the rain, When it snows in yr nose you catch cold in yr brain.
>
> —Allen Ginsberg

The nose has long played a role beyond attracting mates or communing with the gods: since ancient times, it has also served as a vehicle for achieving altered states. The first method of getting high, however, was ingestion. Herodotus observed the Scythians in the fifth century BCE as they inhaled the smoke of burning cannibis seeds. "As it burns, it smokes like incense and the smell of it makes them drunk, just as wine does. As more fruit is thrown on, they get more and more intoxicated until they jump up and start singing and dancing."

Snorting drugs didn't become popular until the discovery and transport of products from the New World, when Columbus noticed American Indians sniffing pulverized tobacco during his second voyage to the Americas in 1494. He returned to Europe with vast stockpiles of the powder, or snuff (the word is derived from *snuffen,* a medieval German term meaning "to inhale deeply"). It quickly became fashionable among the Spanish and French.

English noblemen got hooked after Charles II, who had been in exile in France, returned to London with an enormous supply. Queen Anne so enjoyed the stuff that her entire court took up the habit, and Queen Charlotte, the wife of George III, became so addicted to it that she was nicknamed "Snuffy Charlotte." Her son, George IV, changed his snuff according to the time of day and had a storage room for it in each of his palaces. Commoners, meanwhile, smoked tobacco in pipes because it was a cheaper form of the drug. But in the early eighteenth century they took to snuff, too. It became widely available after kidnapped Spanish seamen ransomed themselves with a huge shipment of snuff. Their British captors then distributed it at ports throughout the country, and it quickly became a fad. Mills and snuff shops became ubiquitous, with more than four hundred stores in London alone.

The drug was so popular throughout the nineteenth century that the period was often called "the Age of Snuff." People carried and exchanged elegant porcelain snuffboxes. Production of the powder far outweighed cumbersome smoking or chewing tobacco. Everybody partook: Charles Darwin used it, Alexander Pope wrote of the "pleasant lift" it gave, and Napoleon sniffed more than seven pounds a month. Doctors used it as a cure-all, prescribing snuff for headaches, insomnia, toothaches, coughs, and colds, and as a general preventative against ailments from tuberculosis to syphilis.

Snuff consumption declined in the mid-1880s, as cigarettes, developed in Spain, became a more popular medium for tobacco. In France, women were particularly fond of the elegant new cigarettes, which were more feminine than pipes and more refined than snuff. Cigarettes soon caught on among both sexes in London, New York, and Boston as sophisticated and "continental."

Tobacco's popularity was soon outweighed by that of cocaine, derived from coca leaves grown in the Andes. Natives of Colombia and Peru had chewed the plant's leaves for centuries; it served as a stimulant and bronchodilator, which was useful for performing physical tasks in such high altitudes. Early explorers returned to Europe with the plant, but it wasn't until the mid-nineteenth century that scientists learned to synthesize it into a more usable powder form. The medical world didn't recognize cocaine's powers as a pain reliever, energizer, aid in sexual performance (as well as a remedy for swollen nasal tissues) until the 1880s, when Sigmund Freud began using it himself. In 1884 he called it a "magical" drug, and was an early advocate of its use as a cure for depression and impotence. He even recommended it for ending morphine addiction (morphine, isolated from opium poppies, became popular in injectable form in the 1850s). Freud changed his mind about the drug's benefits when he spent a night with a morphine-addicted friend trying to wean himself off the opiate with cocaine. The cocaine induced a terrifying night of hallucinations, and afterward Freud cautioned against using it altogether.

In America: Things Go Better with Coke

Americans, far from Vienna, loved cocaine. From the 1850s to the early 1900s, the American elite used a wide variety of drugs, from cocaine, which they snorted, to opium which they smoked. Both drugs were mixed into drinks: cocaine, of course, was an ingredient in Coca-Cola when it was introduced in 1886 (the cocaine was replaced by caffeine in 1903). Snorting cocaine was the preferred medium for most, though, and few were shy about their habit. Thomas Edison praised cocaine's "miraculous" effects, and Sarah Bernhardt gave glowing testimonials, too. In fact, cocaine became a mainstay of the silent film industry—actors and directors found that with it, they had the energy they needed to meet production schedules.

As use spread—people could get cocaine powder in any drugstore—so did concern about the drug's side effects. The drug so powerfully stimulated the brain that addicts often hallucinated, went days without sleeping, and forgot to eat (or bathe). As the drug's effects wear off, depression sets in, leaving the user feeling tired, jumpy, fearful, and anxious—so anxious that addicts desperate for a fix generated a new phrase for the American idiom: "dope fiend." By the turn of the century, American cocaine consumption had grown to an estimated 350,000 (in a population of 76 million), helping to spawn the foundation of the Food and Drug Administration in 1906.

By then, snorting cocaine was a national pastime. Dockworkers along the Mississippi used it to keep energy from flagging, and soon the habit spread throughout river towns from Louisiana to Illinois. Managers of Southern construction camps used cocaine as a means of increasing production among their employees. Western developers found it useful too: overseers in Colorado mining camps found the drug—it had acquired the nickname "snow"—a helpful tool in motivating tired workers, and even supplied it at company stores. Snow found its way to textile mills in the Northeast, where it took the edge off mind-numbing work.

Temperate lawmakers like William Jennings Bryan denounced the drug as a national scourge. Foreign leaders echoed his concerns, but they were troubled for another reason: the United States lacked drug restrictions, which made it a tempting place from which to smuggle. This eventually led to the passage of the Harrison Tax Act of 1914. Far from a prohibition of the drug, its purpose was merely to collect revenue from anyone who imported, manufactured, sold, or dispensed the drug. As a result, the price of cocaine rose significantly. Social critics were outraged.

> As was expected . . . the immediate effects of the Harrison anti-narcotic law were seen in the flocking of drug habitués to hospitals and sanitariums. Sporadic crimes of violence were reported too, due usually to desperate efforts by addicts to obtain drugs . . . The really serious results of this legislation, however, will only appear gradually and will not always be recognized as such. These will be the failures of promising careers, the disrupting of happy families, the commission of crimes which will never be traced to their real cause, and the influx into hospitals to the mentally disordered of many who would otherwise live socially competent lives.[1]

Six years later, Prohibition was decreed, ironically driving addicts hooked on alcohol to cocaine, a more addictive drug. Cocaine was eventually replaced by much cheaper amphetamines in the 1930s, and didn't resurface until well into the 1970s, once the fads of LSD and pot had reached a peak. In the 1980s, cocaine's use grew to epidemic proportions, as imports from South and Central America flooded American cities. It seemed the perfect drug for boom years. Rejuvenated on the club scene in New York, the city that never sleeps, it was as much a staple as coffee bars in today's Seattle.

Its effects, as they had been at the turn of the century, were legion. Hits, depending on their purity, could cost up to several hundred dollars. Even in a fluid economy, the habit ruined fortunes—as well as hearts, brains, and noses. Cocaine stimulates receptors in the cells to release large quantities of endothelin, a chemical that causes blood vessels to contract. Excessive amounts of the chemical can make blood vessels constrict faster and tighter, choking off the heart's blood supply and inducing an immediate heart attack. Seizures are also common. At the height of the epidemic in the mid-1980s, there were one thousand cocaine-induced deaths per year. (Comedian John Belushi died in 1983 of a cocaine and heroin overdose; in 1986, University of Maryland basketball star Len Bias died of a coke-related heart attack.) The deaths did little to stem cocaine use; by 1988, occasional cocaine users numbered around 6 million, and another 4 million addicts snorted weekly or more.

In the materialism of the 1980s, cocaine's rituals were as telling as the quaint snuff porcelain snuffboxes of eighteenth-century England. Cocaine was poured onto a mirror, chopped, separated into "lines," then snorted off a tiny coke spoon, or through rolled-up dollar bills. But snorting took its toll. The powder burns, and over time can wear through the delicate mucous membrane and even perforate the septum. Stevie Nicks, Fleetwood Mac's lead singer, suffered such an indignity: her septum, she has said, eventually bore a hole "big enough to put a gold ring through."

Cocaine use dropped in the early 1990s, only to be replaced by other narcotics. Today, drugs from heroin to glue to ketamine, an animal tranquilizer, are snorted in clubs and school yards, but in truth, the nose is a poor delivery system. Drugs injected into the veins are carried to the brain most rapidly; smoking, in that peculiar race, is second. This is, in part, why the use of crack cocaine—coke that has been cooked into rock form with baking soda or ammonia—became such an epidemic in the late 1980s. (It is also why "huffing," the inhaling of products like spray paint, butane gas, and air freshener, has become so widespread among teenagers.) When smoked, drugs rapidly penetrate the extremely thin lung tissue, which is constructed to allow gases to pass through, and then proceeds via the heart straight to the brain. A drug that is sniffed first has to penetrate the thick mucous membrane of the nose, travel through the blood to the heart, and then return from the heart to the lungs before it can be transported to the brain, resulting in a considerable dilution. The nose's small surface area leaves it an impractical, albeit tidy, way, to get high.

seven

Past Scents: Odor and Memory

Smell is believed to be the most highly developed sense at birth; a crying infant recognizes the unique odor of her mother's breast milk and can be calmed with a piece of her clothing. Without smell, one of life's great pleasures—food—would be scarcely more than fuel, boring and unremarkable. But to understand the importance of smelling, perhaps it's easiest to glimpse life through the experience of one who can't.

Michelle Wood is a trim, active blonde in her mid-fifties with an easy manner, sharp wit, and a hearty laugh—a grown-up tomboy. A professional horse agent, Michelle spends long days outdoors, riding and grooming her charges. She loves the salty air of the Maryland shore near her farm, and cooking—crab-cakes, using local blue crabs, are a particular favorite. Wood takes a special pride in her garden, delighting in the heirloom tomatoes in her vegetable patch and cheery annuals that line her driveway.

At least Wood used to enjoy those things. In the summer of 2000, she got a terrible cold. Chilly and achy on a torpid, hot July weekend, Wood figured she had no ordinary virus. Her nose was so stuffy she couldn't breathe. She couldn't smell, or taste a thing, either. Her family doctor prescribed first one round of antibiotics, then another. Finally, he gave a prescription for steroids, hoping that her post-viral swelling was simply blocking odor molecules from reaching the receptors.

Wood took all the drugs, but nothing helped. Her cold, as it happens, had brought on a malady called *anosmia,* a term forged from Latin and Greek meaning "lack of smell." The disorder is not classified by the Centers for Disease Control, and estimates vary from 4 to 16 million Americans affected. This condition can strike at random, and differs from the gradual fading of the sense of smell experienced by 25 percent of those over the age of 65, called *presbyosmia.* Anosmia can occur after a virus, a bout with sinusitis, years of smoking, repeated chemical exposure, or head injury, which can all destroy the mechanism that fires olfactory information to the brain. In Wood's case, it is likely that the virus attacked odor receptors in such a way that they died off or became unable to bind with odor molecules.

The loss has devastated Wood. She feels, as she puts it, "like I go through the day in a sealed bubble." She still works, traveling frequently to the Netherlands to buy and sell horses. She still tells jokes, and attends to ordinary tasks—cooking, babysitting her baby granddaughter, shopping. But she feels detached, cut off. "Smells have such meaning in my life. As a young girl, I could always weed out boyfriends by whether or not it was OK with them that I smelled like a barn," she says. Now, a dimension is gone from everything. The heady odor of the horses no longer lingers on her boots and hair. Climbing into a bed with clean linens isn't comforting anymore. "I'm there but I'm not," she says. "It sounds crazy to complain when you add it up—I'm walking, breathing, functioning. But that's just it. I'm just functioning."

Researchers believe that the olfactory process helps explain why smell, memories, and feelings are closely intertwined. Odor molecules bind to receptors in the nose, ferrying impulses to the olfactory bulb. The bulb sends messages to the limbic system, the brain's emotional center, and the hippocampus, which is thought to store memory.

For millennia, societies have believed that smell has the power to boost moods and even alter events. In ancient Egypt, embalmers used cedar and myrrh oils to help hasten the mummies' voyage to the afterlife; in ancient Greece, mothers tucked lavender stems in their infants' beds to induce sweet dreams. The scent of a madeleine dipped in lime blossom tea overwhelmed Marcel Proust, launching *The Remembrance of Things Past*. For most, smells usher in powerful sentiments, often for inexplicable reasons.

While science shies away from the term "aromatherapy," some studies show that the effects of pleasant odors are indeed real. Today, Japanese companies circulate the smell of lemon through air-conditioning systems in the morning to help clerical workers stay focused, switching to cedar in the afternoon to boost sagging energy. At London's Heathrow Airport, the scent of pine is sprayed throughout the vast terminals to keep frantic passengers at ease. And doctors at Memorial Sloan-Kettering Cancer Center have found that the vanilla-like aroma of heliotrope, a tiny purple flower with a sweet scent, significantly reduces anxiety for patients undergoing MRI scans.

Yet the ability to smell is perhaps the most underappreciated of all senses. In fact, surveys show that it is the sense a majority of people would be willing to relinquish. Still, current science is showing that the 'Proust phenomenon' is much more powerful than many people realize. Researchers are discovering that smell is vital to memory—and mood.

Wood, for her part, says she doesn't need science to validate her emotional state with research: since becoming anosmic, she has also become profoundly depressed. "What's the mystery?" she says. "Nobody would wonder why I feel this way if I'd suddenly lost my eyesight." Others who have lost their sense of smell report feeling a similar despair. They complain of losing their confidence and becoming paranoid that they stink—many carry around extra stashes of breath mints and deodorants, just in case. They tell of a crushed sense of self—suddenly, they lack clues about how and where they fit in the world. They can't know when the Thanksgiving turkey is done roasting just by smelling, though they've been cooking one for thirty years, or take in the neighbor's freshly cut grass even as he is pushing the mower. They can't smell the skin, or hair, of a loved one. Suddenly, a world of subtlety and nuance becomes altogether flat.

After a while, this sensory isolation becomes self-reinforcing. Many anosmics say that as their illness progresses, they become increasingly secluded. Eating becomes especially unsatisfying. Researchers say that 90 percent of the flavors we enjoy actually comes from a food's aroma: think of a heady sprig of basil, the perfume of an orange, or a thick steak sizzling on the grill. As you take a bite, odors travel up the back of the throat to the nose, where they latch on to odor receptors. (The tongue, meanwhile, savors only sweet, salty, bitter, sour, or a newly discovered, hard-to-describe flavor called *umami,* which can roughly be translated from Japanese as "delicious" or "meaty.")

Since most social gatherings revolve around food, meals are simply new opportunities to be reminded of loss. Wood is a perfect example. Since she became anosmic, Wood, already slim, has shed six pounds—she can't taste anything, so why bother eating? Her withered enjoyment of food mirrors life in her odorless universe: "We went out with friends for crabs not long ago and I became bored very quickly," she says. "Picking crabs is a lot of work and usually a labor of love. But if there is no reward—smell or taste—it is a waste of time." For a time, she and her husband, Wayne, joined friends in restaurants. While others savored their food and wine, Wood would hope in vain for an errant whiff of anything. Before long, though, she realized that she might as well eat burned food at home—she wouldn't notice the difference between it and a twenty-five-dollar steak. She still tries, she says—she loves socializing—but feels that it's senseless to pay for an experience that, in the end, will only leave her bluer than before.

Every so often, Wood will catch a flavor or aroma, but only for a moment. One day she popped a Jolly Rancher candy into her mouth—she can still sometimes distinguish tartness—and was shocked that she could actually taste it. She took it out of her mouth, incredulous, and sniffed the wet red cube as if it were a truffle. "I almost cried, I was so happy," she said. "And of all things, fake fruit!" Instances such as those keep her hopes buoyant, but usually in vain. In December she put up a Christmas tree and lit the bayberry candles she usually enjoys. For a moment, she thought she could smell the candle, but she couldn't be sure. "Was it really the candle, or was I just remembering the smell?" she wonders. The recollection of the pungent pine—or of the loss—is still so palpable, Wood dreamed she smelled a Christmas tree early one May morning, even as lilacs bloomed outside.

When she drives through the Maryland countryside, Wood reminisces about summers she spent on her grandparents' farm: picking strawberries, cranking the handle of the old-fashioned ice cream maker (chocolate was a favorite), or wolfing down her grandmother's chicken and dumplings. On a recent warm day, Wood, her daughter, and granddaughter passed the old farmhouse with their windows down. "There's just something about the air on the Eastern Shore, isn't there? It's different, earthy," her daughter observed. Then she looked at her mother and said, "Oh, God, Mom, I'm sorry. I forgot." Most painful, says Wood, is the link to the past that odors bring. "Maybe some day I will be able to taste and smell my memories again," she says wistfully, "or at least share those experiences through my granddaughter."

At the beginning of her illness, Wood tried to conjure smells. Just after she got sick, she would bury her nose in her husband's neck and take a deep whiff. She could smell—or thought she could smell—his aftershave, the one he'd been using for years. In the beginning, it wasn't so hard: Wood just commanded herself to remember a smell—Wayne, the beach—and she could somehow recall the odor. Now, even that has faded. "I feel like I've lost even my daydreams," she says. "My past—the person I was, and the person I'm trying to remember. You can only go on memory for so long." On a warm day in early summer she was lying on her deck sunbathing. After a few moments, Wood grew restless, and went inside. It just felt too odd. She realized that the fragrance of the sunscreen and the breeze moving off nearby fields was part of what made sunning so pleasant. Without them, she was lost.

Only a handful of researchers actually treat smell disorders, which range from anosmia to "olfactory hallucinations"—smelling odors that simply aren't there. Conditions also include congenital anosmia, being born without the ability to smell, and dysosmia, a condition in which sufferers sense unpleasant smells constantly. Dysosmics report that the stench of garbage, rotten eggs, or a decomposing animal fills their nostrils, driving them to distraction. (Like many anosmics, Wood suffers from occasional dysosmia, and on bad days, she says is overcome by the stink of bacon left in the fridge over a monthlong vacation.) Some, unable to cope with the invasion of horrible odors, commit suicide.

Experts say the small size of their field reflects the fact that smell loss is rarely life-threatening. Every year, however, hundreds of elderly people with diminished smell die in fires or gas leaks others nearby could smell. Many suffer from food poisoning because they can't detect spoilage.

Others, like Wood, suffer much less tangibly. Then there are cases such as Adolfo "Rudy" Coniglio, a pizza chef and restaurant owner from Closter, New Jersey, who in the early 1970s was struck with a heavy cold. A few days later, he returned to work at his pizzeria. But as he settled in among his tomatoes and herbs, he detected a rancid odor. He sniffed the tomatoes: rotten garbage. He sniffed the garlic: rotten garbage. His famous sauce? Rotten garbage. So Coniglio called his suppliers. "Everything's rotten!" he recalls saying. "I was mad, I tell you, I was mad."

"They told me, 'What, you think we'd sell you rotten stuff? You're joking! You're crazy!' It was awful. I couldn't stand it. I went around furious—everything stank. Everyone else says it's fine but I know it's not—it stinks. Even when I cooked in the house it smelled rotten. The vapor from the food, the smells! All I could do was just stand out in the woods with a cigarette. That was the only place I could get comfortable."

Doctors told Coniglio, now eighty-five, that his problem was "in his head"; that he needed tranquilizers; that he needed "rest." In search of respite, he left the country to visit his mother's hometown outside Naples, but the problem remained. "Everything stank. Even in Italy they thought I was crazy. I didn't feel like no crazy guy, but after a while, you got to wonder."

Mainstream medicine has few answers for people like Wood and Coniglio. (Sometimes, smell loss is linked to sinus problems, and can ultimately be treated with steroids. Surgery is also sometimes indicated.) For the most part, there is not even a clear-cut understanding of what causes post-viral anosmia in the first place.

Many anosmics and dysosmics wind up at the Taste and Smell Clinic of Washington, D.C., run by Dr. Robert I. Henkin, a neuroendocrinologist. Coniglio first met Henkin at the

National Institutes of Health in Bethesda, Maryland. Henkin had achieved some renown after he found that tweaking metals in the bloodstream—copper and zinc, largely—could help relieve taste and smell distortions that were the result of viruses or medication. (While hearing and vision tests were commonplace as early as the 1940s, a universal smell test did not exist until 1980. Even its inventor, Dr. Richard Doty, who heads the Smell and Taste Center at the University of Pennsylvania, calls it "an eye test for the nose.")

Henkin tested Coniglio, found his blood serum metals low, and gave him zinc. On the zinc, Coniglio improved dramatically. After his testing, Henkin gave him a placebo with the caveat that he could be on either zinc or a placebo, and asked him to sign a consent form. "But if things get worse," Henkin recalls saying, "call me."

Four months later, Coniglio called. "It all came back," he told Henkin. "Of course!" Henkin said, realizing that the placebo had had no effect. In the meantime, though, Coniglio had sold his business, unable to bear the sensation of rotten garbage any longer. Henkin was horrified. He had just ruined a man's life.

Coniglio, however, shrugs and says it doesn't matter. He had wanted to retire anyway. (He does add, in the next breath, that the net worth of the pizzeria today would be well over a million dollars.) "For a while I didn't know if I'd ever enjoy a bite of veal again! The guy is my hero."

But Henkin isn't a hero to everyone. Doty and Terence Davidson, an ear, nose, and throat doctor at the University of California at San Diego, who also directs the Nasal Dysfunction Clinic there, cautions that none of Henkin's ministrations have received FDA approval—at least for the purpose he employs them. His treatments range from theophylline, an asthma medicine he says helps stimulate enzyme production needed to help olfactory receptors work properly in anosmics, to anticonvulsants and antipsychotic medications for those who suffer from distortions. (Wood visited Henkin, and says that theophylline he prescribed made her anxious and gave her heart palpitations. She stopped the treatment and never went back.)

On the other hand, patients with smell disorders have few places to turn. In the medical community, anosmia is an "orphan disease." There is little research in the area, and drug companies rarely fund the development of medications upon which they can't guarantee profit.

Smell loss is so obscure to many physicians that they simply dismiss it. Those who do seek help are told by their doctors that they should simply "learn to live" with their problem. Furthermore, insurance companies do not often cover treatment for anosmia; consultations and evaluations can cost several hundred, or even several thousand dollars. "There isn't widespread knowledge about a lot of these conditions," says Doty. "And it's not all that hard to understand, really: some diseases have decades of research and understanding—heart problems, vision problems. We're just not in that same place yet."

Many with smell disorders often run through a long string of specialists—ENTs, neurologists, even psychiatrists. "The system doesn't treat them very well," says Doty. "It's like, 'What? I've got a room full of patients with tumors and you're complaining that you can't smell your coffee? Get out of here!' There's a lot of that, I'm sorry to say. But these people are suffering. They have lost a great deal. We spend a day evaluating these people. We can't always help everyone—I'll be honest. And for those we can't, we try to at least help them get closure on what is realistic, and what isn't."

Almost a year after her virus, Wood has high hopes that she will get her sense of smell back on her own; occasionally, odors creep into her nostrils, if only fleetingly. One brilliant spring day she unlocked her trunk to retrieve some horse strapping. As the door flew open her nostrils filled with the smell of warm leather. "I breathed it in and realized it wasn't just a memory, it was the real thing," she says. "You'd have thought I won the lottery."

Anosmics like Wood, who have lost their sense of smell to a virus, often experience a slow recovery, Doty says. Since olfactory neurons regenerate every month or so, one would think that those who lose their sense of smell would recover—or at least notice improvement—within weeks after the initial loss. Instead, it takes months, sometimes even years, for ex-smokers to regain much of their sense of smell. Smokers, Doty points out, suffer mild damage, and are still able to identify many odors. "Obviously, most people with smell loss haven't had their olfactory abilities tested before becoming injured or ill," he says, "so it's impossible to compare." Those whose sense of smell is not completely gone, he says, have a better chance of improvement.

For many years scientists could not explain why we remember smells. Olfactory neurons are constantly being replaced. New neurons have to form new synapses, even when smelling an odor you've smelled daily for twenty years. If that is so, why would the brain remember odor? Buck and Axel, the Columbia researchers who discovered the genes responsible for smelling, hypothesize that memories survive because the axons of the neurons corresponding to various receptors always go to the same place in the brain.

For her part, Wood gets by. "It's really easy to feel sorry for myself," she says. "But then one morning I'll get a whiff of the garbage that didn't get emptied the night before—and I'll just be so thrilled I could scream. Garbage? Rotting banana peels? Wayne's aftershave and horse manure still elude me, and those are two things I long to smell. It's an odd combination, but that is the stuff my life is made of, and I need it back to feel complete."

According to recent studies, brain scans using functional magnetic reasoning imaging (fMRI) among those with a normal ability to smell showed that the brain "activated" when exposed to three different chemical odors. Scans of those who could not smell revealed little or no activation of the brain, including the hippocampus and limbic system. Wood is depressed to hear it. "Good God, I wonder what's happening to *my* brain," she says, laughing grimly. She spurns any suggestion of pharmaceutical help, like Prozac, to "adjust" to her disorder. No pill can help her distinguish grilled salmon from Starkist tuna, so what's the point? Still, some measure of human enjoyment has been drained out of everything—eating, her work, sex. "It's not because I have a lack of interest in my husband," she says. "It's because I can't smell him." Experts draw the same conclusion. Anosmics often have a diminished sex drive. Rats whose olfactory bulbs are removed become disinterested in mating.

* * *

What of the emotional lives of people who cannot detect the scents around them? Are they deprived of full happiness—or even of their capacity to remember? Rachel Herz, the professor of psychology at Brown, says scientists are only beginning to realize the connection among the three. "People who lose their sense of smell for extended periods generally feel as if their emotional lives are blunted," says Herz. "But we can't say for sure why. This is total conjecture, and it needs to be studied, but it is possible that without a constant stream of sensory input the limbic areas of anosmics actually do start to atrophy."

The ability to smell adds an extra dimension to the sipping of wine, a luxuriant bubble bath, a hike outdoors. But beyond the scents themselves, researchers say that smells can help to elicit memories in a way that other sensory stimuli can not. Scent is often said to be the most powerful conduit to memory. "It's much more subtle than that," says Herz, a small woman with green eyes and long auburn hair. When an odor triggers a memory, she says, the emotions that accompany the recollection are often more intense than memories spurred by, say, music or a photograph. "Because your feelings are so strong, they may make you feel more confident that what you are recalling is especially accurate," she adds.

Memory, after all, is more than just an accurate representation of a past event, or a simple recall of facts. Brain scans taken as a person remembers Paris as the capital of France differ greatly from those of the same person reminiscing about a first trip there. Not all memory is emotional, of course, but the recollection of specific episodes often is. "People associate past events with feelings—where they were, what they were eating, who said what," Herz says. "Those feelings can range from a kind of hazy nostalgia to intense emotional responses." That's why vanilla-scented pipe smoke—the very kind of tobacco your grandfather smoked before he died—may make you tear up. Or why you continue to buy the brand of dish soap your favorite aunt used—it reminds you of warm evenings spent at her house.

In a study Herz conducted with Jonathan Schooler, a psychologist at the University of Pittsburgh, subjects were prompted with words for an item—in this case, Coppertone suntan lotion, Crayola crayons, Johnson & Johnson's baby powder, Vick's Vapo-Rub, and Play-Doh. (Herz and Schooler thought these products would be more likely to trigger childhood memories and not recent ones.) The subjects were asked to search for a memory associated with the word for the item, and to rate it for the feelings it elicited, how vividly or in how much detail they could recall the event. After a pause, the researchers presented the sight or the smell of the item, and asked subjects again to think about and rate the episode they had just described. Vividness and specificity didn't vary in the ratings the second time, the researchers found. But when odor was the cue, subjects recalled the same incident with much more intense emotion. "They really felt as if they had been transported back to the original event," Herz says.

In another experiment, Herz showed people a series of emotionally evocative paintings. At the same time, the subjects were exposed to another sensory cue—an orange, for example—in different ways. Some saw an orange. Others were given an orange to touch, heard the word 'orange' or smelled the fruit. Two days later, the subjects were given their sensory cue once again and asked to recall the painting. The sensory cues were all about the same in eliciting recall of the picture, but those exposed to smell not only remembered the painting, but felt a "flood" of emotional responses to it as well, Herz says. This insight, she believes, helps explain a lot about smell and memory. "It's the level of emotional intensity that makes it seem as if a whiff of a former lover's cologne somehow reminds you of him better than just looking at a picture of him," she says.

Smell enhances life's pleasures, but the reverse is also true. Researchers are exploring whether impaired smell in the elderly, among chemical workers and smokers, might affect mood disorders as well. Duke University researcher Susan Schiffman suggests that smell loss in the elderly can lead to unwanted weight loss and deepen depression, which affects about 10 percent of those over fifty-five.

Not surprisingly, there is also a connection between scent and negative emotions. In a project conducted with Gisela Epple at the Monell Chemical Senses Center in Philadelphia, Herz and Epple asked the children to complete a lap-sized maze that was in fact impossible to solve. At the same time, the researchers piped an unusual odor into the room. Twenty minutes later, they gave the children an easy assignment in another room filled with the same odor, a different odor, or no odor at all. The children who were given the simple task in the room scented with the same odor as that of the maze room did significantly worse than the other boys and girls. Meanwhile, there was no difference in performance among children in either the no-odor or different-odor rooms. The researchers concluded that the performance rates were not due to the presence of maze odor, but to the specific associations the first odor triggered—the experience of failure.

So, if you flunked your first exam in a classroom freshly scrubbed with bleach, you might be at risk of repeating your performance in another recently Cloroxed room. Scientists believe that odors not only affect our memories—but how we respond to them. Researchers in France found that newborns prefer the odors of breast milk and amniotic fluid to those of vanilla and baby formula. Certainly, those preferences don't last long, but events certainly help shape whether we like an odor or not. Trygg Engen, a professor emeritus of psychology at Brown, wrote that a person's affinity for or dislike of an odor is determined by his first exposure to it, and is bound forever to the joy, hilarity, or sadness of the experience. Anecdotal evidence of this is everywhere. One woman I know can't bear the aroma of marinara sauce; it was what her mother was cooking the night her father had a massive heart attack and died. For years, the scent of Juicy Fruit or Doublemint gum repulsed my sister, who believed they made her nauseous. In fact, it was, and is, what our father always chews, and he used to go through a whole pack on the winding road through the Cascade Mountains as we went for weekend ski trips. It was only years later, as a driving parent herself, that she realized the *road* made her sick, not the gum.

eight

The Sinus Business

It feels like the most common of maladies, a cold, with its runny, stuffy nose, headache, and general malaise. Most people can treat their symptoms with nothing more than a day or two off from work, a box of Kleenex, and an extra carton of Tropicana. But for the estimated 37 million Americans who suffer from chronic sinusitis, the scratchy throat that marks the beginning of a simple respiratory virus is met with utter dread. It is the start of a cycle that typically leads to visits to the doctor (if not the hospital), rounds of antibiotics, and weeks of misery. For reasons that science has not been able to fully understand, the sinuses of these people never function properly, and collapse at the onset of a case of sniffles. The gaping chasm between what medicine can provide, and what human beings need, can be witnessed dramatically in their sufferings.

The disease has stumped doctors by its sheer numbers. In 1982, the National Center for Health Statistics reported 27 million cases of chronic sinusitis; by 1993, that figure had spiked to 37 million, affecting roughly one in seven Americans. According to the NCHS, chronic sinusitis is the most common long-term disease in the nation, overtaking asthma, arthritis, hypertension, and back pain. More patients are reporting symptoms of the disease than ever before—NCHS records what patients seek treatment for, not doctors' diagnoses—and physicians, armed with new machines like CT scans and endoscopes, are finding increased cases as well. Neither the rise in patients' symptoms nor improved tools for finding the disease can explain the dramatic growth.

Chronic sinusitis differs vastly from allergic rhinitis, but people often confuse the two. When pollen collides with cells called "mast cells" just beneath the lining of the nose, eyes, or respiratory tract of susceptible people, the immune system goes into overdrive. Once disturbed, mast cells go into high gear, producing some 15 inflammatory molecules. One, called histamine, leaps to defend the body against the invader. Histamines attach to receptors on nerve cells, which then lodge on receptors in nearby blood vessels. The vessels become porous and seep with fluid that becomes the most notable feature of an allergic reaction: the ticklish, runny nose, the leaky eyes and scratchy throat. Antihistamine drugs work for allergies, but not for chronic sinusitis. The active ingredient in antihistamines is a molecule that fits the histamine receptor, thwarting the signal that causes the drippy signals. Often, patients and doctors alike confuse the symptoms, but antihistamines do nothing for patients with chronic sinusitis, and can sometimes make the patient worse by drying up the tissues.

The sinuses themselves are something of an enigma. Experts don't agree on their original function. One theory suggests that they evolved over time, in order to make our heads lighter, once we began walking upright; some say they exist to cool, heat, and humidify the air that goes into our lungs. Others say that they're there to equalize barometric pressure changes. One thing is certain: sinuses are the seat of the voice's resonance, acting for a singer much like the chambers of a violin.

The nose, sinuses, and their interconnecting ducts, no bigger than 2 millimeters in diameter, are lined with a mucous membrane, which, when healthy, resembles the lining of the inside of the mouth. The mucous membrane produces between a pint and a quart of mucus a day, which is flushed out of the sinuses, a grapelike cluster of cavities, by thousands of tiny *cilia,* or hairlike filaments. They help pass pollen, bacteria, and viruses out of the nose at the back of the throat. (Many bigger particles, such as dust and pollutants are trapped by visible nasal hairs, and eventually gets blown out.) The mucus, along with any viruses and bacteria it may carry, gets swallowed and then dissolved by stomach acids. These functions go largely unnoticed by most people, but for those who suffer from chronic sinusitis, even breathing and swallowing are conspicuous.

In the late 1990s, a small band of researchers at Minnesota's Mayo Clinic began to question the conventional treatments of the thousands of patients who flooded the clinic's ear, nose, and throat department. They are led by a young German doctor named Jens Ponikau.

A tall, bearish man with a broad face, Ponikau has a crushing handshake, and thinning, close-cropped blond hair. His English is marked by the year—1983—he spent in Minnesota as a foreign exchange student; he says "Awesome!" or "Neat!" when something strikes his fancy. He smiles broadly and claps colleagues on the back as he rushes past them in gurney-filled hallways—a jolly German. Little about his appearance or demeanor reveals that he is a revolutionary in his field, the rarefied world of ear, nose, and throat surgeons.

Born in Halle, East Germany, in 1966, Ponikau's earliest memory is the night he and his mother stowed themselves in a trunk to steal across the country's western border. Acting on a tip from their smugglers, the country's security agents, the Stasi, raided the car. "It is quite something, as a child, to look up and see three AK-47s pointed at your face," he says. His parents were jailed for treason. Young Jens, an only child, spent two years in the care of his grandmother.

Eventually, in 1978, the West German government ransomed a group of highly trained political prisoners; Jens's father, a doctor, and his mother, a teacher, were among them. The family settled in Hamburg, where Jens aspired to be a doctor. In his last year of medical school, he spent some time at the Mayo Clinic in Rochester, Minnesota. There, he met some ear, nose, and throat surgeons with whom he felt an immediate affinity—his father is also an ENT—and he planned to enter the field himself.

For Ponikau the subspecialty is not just the family profession, it is fascinating in itself. The ears, nose, throat, and sinuses, he says, are central to the human experience: hearing, smelling, breathing, talking. Ponikau was especially interested in chronic sinus disease and its precipitous rise. Back at Mayo as a resident, Ponikau found himself under the tutelage of Eugene Kern, a mustachioed New Jerseyan whose skill as one of the clinic's sinus surgeons had earned him a national reputation. Every year, Kern assigns his residents a project. Ponikau's was to look at fungi in the sinuses. He remembers the day he got the assignment—it was a wintry morning in Rochester—because it would touch off his research career.

In sinus surgery, the excised tissue is sent to the lab to be analyzed for bacteria and irregular cells. In rare cases, 4 to 5 percent, patients were found to have a condition called allergic fungal sinusitis, or AFS. Strangely, though, a great many people with chronic sinusitis seemed to have fungal growths lodged in their sinuses. But cultures, didn't usually find anything. What doctors could see—or thought they could see—was not confirmed by lab results.

And yet fungi, like bacteria, are everywhere, Ponikau reflected one evening as he hunched over his microscope and slides. Spores flourish in the air—inside walls, shower stalls, wet basements, under Formica countertops, in the soil; they even are the food we eat—mushrooms, of course, being the perfect example. It seemed perfectly likely that fungi existed in the sinuses as well—the perfect moist, warm incubator.

Ponikau was particularly mystified by the symptoms of chronic sinusitis patients. Their complaints were numerous, and real: persistent congestion, a runny nose, headaches, facial pain, fevers, and other flu-like symptoms that can last for months. Often, like Ponikau himself, they suffer from asthma as well. "Put together, the disease seemed systemic," he says. "There is rarely just a localized problem; sometimes patients come in and say, 'I know the infection is back because my asthma has flared up again.' "

Ponikau trudged back and forth between the lab and Mayo's vast waiting room, equipped with television sets, board games, and an adjacent espresso stand, thinking about the patients symptoms of those who shuffled forlornly into Gene Kern's office. At best, chronic sinusitis patients complain of low energy. But for most, the illness dominates everything, forcing absences from work, school, and social functions. Visits to the doctor's office and the pharmacy become as frequent as trips to the supermarket. Unlike chronic fatigue syndrome or fibromyalgia, maladies the medical community disputes, chronic sinusitis has quantifiable measurements: CT scans reveal diseased tissue, and doctors can see swollen, inflamed mucosa and infected mucus. The disease is also responsible for 73 million lost workdays, according to a 1990 Harvard study. It is rarely fatal, but its complications—meningitis, encephalitis, or pneumonia—can be. "There isn't a single doctor who deals with chronic cases of this disease who doesn't want to throw up their hands," Ponikau says.

Unlike the heart, lungs, kidneys, and immune system, whose function can be evaluated with simple tests, few clinics are equipped with tools to measure sinus function. (Mayo is one of the few places in the United States with a nasal physiology lab. Doctors there can determine how well a microscopic sample of cilia passes along mucus.) Elsewhere, the only test is whether or not patients feel better after one or more types of treatment.

Surgery is a last resort. Many who exceed the ken of their own physicians wind up at Mayo. Most of this group, by this time, has had multiple surgeries—one patient Kern saw had had twenty-five. They also go to remarkable lengths to try to improve their condition themselves: selling a house they think makes them sick, moving across the country to a different climate, buying special machines to change the ions in a room, taking allergy shots for years. On the advice of self-help books, health store clerks, and Internet doctors, patients drink vinegar, swallow acidophilus pills, drop oregano oil under their tongues. "I've talked to patients who go on diets that won't allow carbohydrates or fruit—no fruit! This is religion, not science!" Ponikau says.

Whatever is causing the problem, one thing is for certain: the disease translates into big business, especially for drug companies. The 1990 Harvard study found that the average drug costs for patients with chronic sinusitis totaled $1,220 per patient, per year. Drugs to treat the disease include antibiotics, antihistamines, decongestants, and steroids, either in nasal sprays or pills, or both. That works out to some $45 billion on drugs a year. The same study found that the cost of surgery averaged nearly $7,000 per patient; there are some 250,000 sinus surgeries performed in the United States per year. The Mayo Clinic's business office did a study that found actual costs of sinus surgery were much higher. Once blood tests, anesthesia, and hospital costs were included, the figure neared $12,000. That would make a national total of surgical and hospital costs close to $3 billion. None of those figures take into account the sums Americans spend on alternative or over-the-counter (OTC) care. One study, released at a San Antonio conference on sinus disease, estimated that Americans spend $15 billion on OTC medicine and alternative therapies like acupuncture treating themselves for sinusitis. According to the CDC, in 1998 chronic sinusitis was responsible for 12 million doctor's visits and 1.2 million hospital admissions.

The economic implications are obviously far-reaching, and the rewards for the drug companies are great. According to Kline and Company, an independent market analyst firm in New Jersey, the consumers spent $200 million on OTC nasal drops, inhalants, and sprays, up 10 percent over 1999; OTC sinus medications, such as Sudafed and Tylenol Sinus, accounted for another $150 million. Pharmaceutical companies clearly recognize the size of the epidemic; one study showed that in the mid-1990's, the commercials aired most often during winter on the largest networks, ABC, NBC, and CBS, featured cold and sinus remedies, with combined revenue for the ads averaging $600 million yearly. A book on sinus disease called "Sinus Survival: The Holistic Medical Treatment for Sinusitis, Allergies, and Colds" has sold more than 300,000 copies.

Acute sinusitis, which generally appears after a bad cold or a prolonged allergy season, is easily treated with antibiotics. But chronic sinusitis is much more elusive. For years, doctors have attributed it to a variety of causes: allergies to pollen, mold, or animals; sensitivity to pollution and chemicals; bacteria trapped in the sinuses. Genetic disorders can also play a role. The only constant is the sufferer's diseased, damaged epithelium, which cannot adequately move mucus out of the sinuses. Constant inflammation of the sinus lining can also lead to polyps, small, benign growths that further block the interconnecting passageways.

In addition to medication, doctors have also sought to treat the disease with surgery, by creating larger ducts, removing congenital blockages or scraping away tissue deadened by repeated infections. Surgery for repeated sinus infections was advocated as early as the nineteenth century, but it was a risky and unwieldy procedure. Doctors could only gain access to the sinuses by cutting incisions in the roof of the mouth, or, more drastically, on either side of the bridge of the nose beneath the eyebrow.

In the mid-1980s, Dr. David Kennedy, chairman of the otorhinolaryngology department at the University of Pennsylvania School of Medicine, introduced American doctors to the use of endoscopes in sinus surgery, a technique he had learned from a surgeon in Austria named Heinz Stammberger. Endoscopes, small telescopes attached to a video monitor, allowed surgeons entryway through the nostrils. These gave a magnified view of their tiny operating field. Endoscopes allowed sinus surgery to become commonplace, with large ENT practices advertising happy, easy-breathing "customers" in newspapers and yellow pages everywhere.

Quite literally, the technology opened up a whole new surgical field. Endoscopes allowed easy removal of tissue that doctors suspected was to blame for chronic infections. Surgeons took out turbinates, stripped away membranes, and created new, larger ducts in the floor of the sinuses.

But by the mid-1990s, doctors began to question the wisdom of such aggressive procedures. While many postoperative patients enjoyed a level of health they had never experienced, many others began showing up at Kern's office. Many complained that their illness worsened *after* the surgery, and that their symptoms had increased. Not only did they have chronic infections, they now suffered from bloody noses and had extensive breathing problems. They were also very depressed.

Soon Kern noticed that 157 of his new patients had another thing in common: they had all had their turbinates removed. Looking at their CT scans one day, a Swedish colleague, Monica Stenquist, even came up with the name for the new condition: empty nose syndrome. Scans of normal, healthy sinuses show the turbinates curling from the outside of the nasal wall into the air, like miniature fiddleheads. In the X-rays the doctors were reviewing, all lacked the bony structures.

Humans, in fact, need their turbinates: they are essential for humidifying the nose and keeping disease from traveling to the frontal sinuses, which are separated from the brain by the thinnest of bones. More than 160,000 people had turbinate tissue removed in 1996, the last year for which the Centers for Disease Control and Prevention has statistics. (Kern believes the true number is far higher and that it is rising.) Other radical procedures had proved equally disappointing. The new, bigger ducts created by surgery went unrecognized by the cilia, which just kept using the ones nature had made. Often, stripping the sinus membrane permanently damaged cilia. Technology had given well-trained physicians a boost, but in the wrong hands, it had become a destructive tool.

Even in the best of circumstances, sinus surgery is a difficult procedure. The surgeon, generally a man (83 percent of ENTs are male), must possess fine motor skills comparable to those of a concert pianist. As in art, there are Picassos, and there are journeymen, and the skill required to remove damaged tissue—but leave the delicate cilia intact—is refined indeed.

For the private practitioner, sinus surgery can be a lucrative subspecialty, bringing in as much as $500,000 a year. Doing the surgery itself, many say, is "fun." Part of that fun derives from the arsenal of high-tech medical toys many doctors have at their disposal: fiber optics and micro-cameras, to better peer into the dark, cramped space; giant monitors that magnify the field of action; and three-dimensional CT scans. Some operating rooms are even outfitted with $160,000 global positioning systems that beep when the surgeon is too far off target—or too close to severing a nerve.

In fact, the only thing about which sinus surgery cannot boast is success. In addition to complications from empty nose syndrome, thousands of other Americans undergo repeat procedures every year. While there are no certain figures, the revision rate on sinus surgery is estimated to be anywhere from 20 to 50 percent. One reason the figure is so difficult to track is because patients often switch doctors once one surgery fails them, only to repeat the process elsewhere.

The Harvard study did show that patients require fewer drugs after surgery, but it is possible that those figures disproportionately reflect patients with congenital blockages or too-small ducts.

At any rate, the "simple ambulatory procedure," as the surgery is often described, is anything but. For a painful postoperative week, the patient grits through swelling, bleeding, and the immense pain of having had bone and tissue removed from one of the most nerve-dense regions of the body. Several days after the surgery comes the sensation—and indignity—of having wads of bloody cotton packing extracted from one's nasal cavities: "Close your eyes now, you'll feel a little tug!" (As a veteran of both childbirth without anesthesia and multiple sinus surgeries, I'd choose labor any day.)

"If surgery cured people, we'd all be happy," Ponikau says. "But clearly, it only temporarily helps most patients."

At the beginning of Ponikau's tenure at Mayo, the whole department debated the disease. Many times, it felt like the team was "sitting" on an answer, particularly when postsurgical patients would return just as sick as before. The department held meetings on difficult cases to determine the "right" surgical plan, one that would solve the poor patient's problems forever. But six months later, Ponikau would see him or her looking dejected in the waiting room. "You felt heartbroken just to see them, and know that you'd failed them," he says. "Sorry, Doc, I'm back," they'd say. And you'd say, "What are you sorry for? We're the ones who are failing *you!*"

In many cases, it was easy to lay blame elsewhere. One busy afternoon Gene Kern rushed into the examination room to see a patient. The man had several surgeries behind him, and clutched a new CT scan showing disease in every cavity. "Well, this poor guy had some inexperienced doctor do his surgery," Kern thought to himself.

"When did you have your last procedure?" Kern asked the man.

"Nine months ago, Doctor."

"I see," Kern said. "Who did the surgery?"

The man cleared his throat and flinched for a moment before looking Kern in the eye.

"You, Dr. Kern. You did it last May."

Stubborn cases eventually made Kern question not only the mechanism of the illness, but its accepted treatment. Was constant surgery a crude—and brief—way to ameliorate symptoms? "As a surgeon, your main goal is to help these patients feel better," Kern says. "But when you looked at this objectively, a lot of what we were doing just wasn't really helping."

The department often debated the cause of the disease, at lunch in the cafeteria or at weekend barbecues. At Mayo, where the dozen or so clinic buildings are connected by a series of underground passageways referred to as "the subway," there is an attitude of surprising affection and collegiality. In many other labs, breakthroughs are carefully guarded from everyone except research partners. But at the elite fraternity of Mayo, ideas are currency. If you have an interesting theory, you pick up the phone, and within hours or even minutes have the second, third, and fourth opinions of a staff pathologist, an immunologist, an epidemiologist.

One day Ponikau realized that maybe the department was looking at the wrong thing. Maybe it wasn't the tissue, or a bacterium, or a virus, that was causing the problem: maybe it was the body's own immune system. Together, Ponikau and a surgical pathologist, Tom Gaffey, began looking at patients' tissue under an electron microscope, which can magnify cells up to 200,000 times. They saw something unusual: clusters of white blood cells called eosinophils clumped near the surface of the membrane.

"For a while, we forgot about the fungi," Ponikau says. "My whole life was eosinophils."

Eosinophils normally exist in the bloodstream in minute numbers—the range is anywhere from 1/200 to 1/2000 of white blood cells. They are incompletely understood, but their chief role is thought to be in combating parasites. Eosinophils contain several dense granules comprised of powerful proteins, which they use to destroy invaders. "Stupid warrior cells," Ponikau calls them. "Rambo-like." Gaffey and Ponikau experimented with fluorescent staining techniques so they could better locate eosinophils. Once they had perfected the new stains, they noticed something unusual: the immune warriors were literally lined up in the basal cell wall of the mucous membrane—the meeting place between the body and the outside world. Their concentration had jumped from an average of .01 percent in the blood to 20 percent in the tissue—a 2,000 percent increase.

Until that moment, the conventional wisdom in medicine had been that viruses, bacteria, or chemicals irritate or infect the lining of the sinus. Get rid of the invader, end the problem. Instead, Ponikau realized, the key to the disease might be in what the body was producing itself: the mucus produced by the membrane to moisten and protect it. For surgeons dealing with the disease, mucus was generally an irritation—something to be suctioned out and discarded. But in fact, Ponikau thought, the mucus likely had its own story to tell. The eosinophils had accumulated at the membrane wall and were even entering the membrane itself. Perhaps, he thought, they were even going into the mucus. But why? Clearly, the body had deployed them to destroy something, much as bacteria are attacked by neutrophils, another type of white blood cell. If Ponikau's theory was right, chronic sinusitis patients should be disproportionately those people whose phlegm was full of esosinophils. In order to follow up on his hunch, Ponikau needed vast quantities of mucus.

When surgeons set about operating, the first thing they do is clear the mucus—it blocks the tissue they need to reach. Ponikau had to convince first Kern and David Sherris, one of Kern's partners, that he needed the mucus. But to do that required a reversal of operating room procedure as it had been practiced for decades. Ponikau's strange request for a waste product required the support of some thirty staff members, from scrubs nurses to lab technicians to pathologists, all of whom were accustomed instead to preserving tiny bits of diseased, excised flesh, but not mucus.

Operating room procedure was not the only hurdle. Once the mucus was extracted, Ponikau had to ensure the safety of his samples up to the pathology lab, several floors away. Running through tunnels and hallways in his teal scrubs and clogs, mask dangling from his neck, Ponikau cut an amusing figure. "Don't drop the mucus! Don't drop the mucus!" he would shout as he jogged through the gleaming white corridors.

"It was as if we were carrying some sort of ancient, precious glass," says Karmen McGill, a surgical technician who was among the first of Ponikau's converts. "It was pretty funny. People would look at you as if you had an organ for transplant or something, and here it was just mucus on a slide."

It was a complex process, and it aroused deep suspicions. A pathologist once called him, irate at the strange specimen that had appeared in his lab. "Somebody delivered snot up here!" he shouted. "What the hell are you people thinking?"

Gaffey began reflecting on Ponikau's theory in his off-hours. A tall, affable man with jet-black hair, Gaffey is fond of murder mysteries and jigsaw puzzles, and grasped the possible significance of Ponikau's speculation immediately. For years, Gaffey had sent lab reports back on sinus surgery patients which read: "Evidence of chronic inflammation. No cancer. Tissue marked by 10 percent eosinophilic penetration." Looking back, Gaffey saw a pattern: the higher the penetration of eosinophils, the more intensely the patient suffered from the disease.

Gaffey and Ponikau began to search the mucus under the electron microscope, and found clusters of eosinophils there. Maybe a parasite was in the mucus, Ponikau announced to his colleagues. Sherris laughed out loud. "Now the German's really gone nuts," he said. "Send him back home."

As a control, Ponikau had collected mucus from healthy medical students who had no allergies—no nasal symptoms whatsoever. To their surprise, Gaffey and Ponikau found the common fungi in both patients and healthy controls. At first, it looked like the end of his theory. But then Gaffey started to notice something strange. In the mucus of healthy patients, the fungi were characterized by their branch-like appearance (under a microscope, they look like saguaro cacti). "Textbook," Gaffey says. One afternoon, Gaffey had a "Eureka moment." Sinusitis patients' fungi, under the microscope, looked misshapen and odd-looking by comparison. It looked "destroyed, beat up," he says, "as if it were sitting in a cloud of toxic protein."

Suddenly, Ponikau's fungi assignment for Kern looked like scientific fate. When the two scientists realized that the eosinophils were forming clusters around the fungi, Ponikau wondered if his early, coincidental interest might have an effect on the understanding of the disease. Everyone had fungi in their mucus, but only the sinusitis patients had eosinophils there. Finally, Gaffey and Ponikau understood the meaning of their findings. The eosinophils were crossing the blood barrier—literally leaving the bloodstream in order to search out the fungi and destroy it. The eosinophils were releasing MPB, major basic protein, which is known to be toxic to parasites. Unfortunately, MBP is also toxic to the delicate mucous membrane, leaving it damaged and ulcerated, even destroying cilia. It wreaks havoc on the whole system. Bacteria weren't the problem—studies had long established that they were as rampant in healthy people as in sinusitis patients. But if the tissue gets inflamed or ulcerated—which it does when the MBP is released—the bacteria can penetrate and invade the body. Bacterial infections are common in chronic sinusitis patients, and are known as acute exacerbations of the disease. But the bacteria's presence obscured the fungi. Bacteria are easy to culture, and, absent antibiotic resistance, are simple enough to treat.

Ponikau points to an ugly, swollen gash on his left hand, the result of a scrape with his garage door. "It makes perfect sense. The skin on my right hand is clear—no red, infected-looking scabs. Why? Because the skin is protecting the tissue from all the bacteria that normally live on top of it." Once the skin is cut, bacteria can rush in and multiply. This, he says, is exactly what happens in patients with chronic sinusitis, again and again. The reaction to the fungi is like a cut that never heals—it allows the bacteria to invade.

While the clinic's findings had yet to be published, Ponikau felt confident enough about the fungal link to present his theory at a 1998 conference of rhinologists and allergists in Vienna. The news was met with mixed reactions.

Heinz Stammberger, an Austrian ENT who, along with his mentor, Walter Messerklinger, is largely credited with pioneering endoscopic sinus surgery, listened intently. Stammberger himself had written his doctoral dissertation on fungal sinusitis, and looked on in stunned amazement.

Stammberger, a tall, slender man with a crisp British-German accent, was both surprised and disappointed by the discovery. His own team of physicians at the ENT University Hospital in Graz, Austria, had been looking at fungal disease for years. "I felt that we should've come up with the diagnosis. But we simply didn't have the idea Jens had. He had the right concept, and followed through." Stammberger was almost as astonished by the reaction of other doctors, many of whom did not welcome the new theory. At many meetings over the next few years, people were aggressive and often downright rude. "They simply did not want to hear this," he says. "The findings could spell radically new treatments."

At one conference, a young ENT marched up to Ponikau, arms folded across his chest. He felt that his surgical skills, hard won after years of training, were in danger of being made obsolete. "I am thirty-three years old, and have been waiting for the day I could practice medicine in my subspecialty now for fourteen years. I worked part-time in college, put myself through med school, and, after my internship and residency, am almost $200,000 in debt. I am about to get married. I want to pay off my bills, buy a house. But what you've just shown us might obviate all my hard work. Do you realize that?"

Others took the results—published in the Mayo Clinic's journal, *The Mayo Clinic Proceedings,* in 1999—less personally. Hannes Braun, an ENT at the Graz Hospital, believed Ponikau, but thought the problem was environmental. Braun had always thought that fungal growth was an American predicament, somehow tied to contaminated air-conditioning filters. "I was interested, sure, but I didn't think these results could be duplicated elsewhere—most of the rest of the world doesn't have air-conditioning."

Stammberger and Braun replicated the study with ninety patients, many of whom had never drawn a breath of air-conditioned air in their lives. As controls, they used healthy doctors and medical students. They, too, found eosinophils in the tissue and mucus. They also found fungi in the mucus of more than 92 percent of their subjects. The team could no longer say it was a function of American air conditioners. "Ponikau was right," Braun concedes.

Walter Buzina, a microbiologist on staff at the clinic, showed Stammberger his lab results: a total of 283 positive cultures had grown—an average of 3.2 organisms per patient. As a mushroom lover in the Central European tradition—he hunts for them on weekends—Buzina decided to let the specimens take their course. Some weeks later, he summoned the team into the lab for a viewing: enormous mushrooms had grown from the swabs from subjects' noses. Specimens ranged from huge blooms of *cladosporium,* the black growths typically found in dank showers, to *penicillium,* the soft blue-green mold that appears in forgotten bread bags, to the graceful gills of *tricholoma,* prized by mushroom lovers for its delicate flavor and aroma. "We couldn't quite believe these growths ourselves," Buzina says. "It was quite amusing."

Buzina presented his findings—on slides—at a conference of nose experts in Washington in September, 2000,[1] and the mushroom harvest got a big laugh. Ponikau, for his part, addressed the crowd by saying that the new theory was a "major shift in the paradigm" of understanding sinus disease. Dr. David Kennedy followed up with a talk concluding that many of the field's assumptions about sinuses now appear as "a collection of facts, theories, and fetishes," obscuring multiple causes for the disease—heredity, allergies, pollutants, and stress. Studies from Finland to Turkey link the disease to pollution. Kennedy recommended more studies to evaluate genetic or individual factors, as well as the role of bone inflammation in the disease.

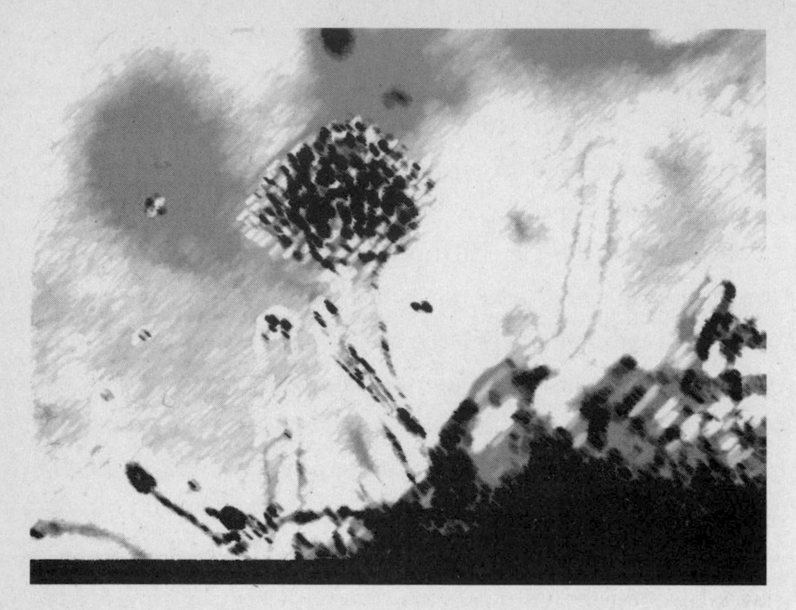

Aspergillus, one of the nose's most common fungi. (Photo courtesy of Dr. Walter Buzina.)

Kennedy admits that surgery is not a panacea, but he feels that "it does help resolve symptoms over time." He adds that it can take up to two years to fully recover from the postoperative inflammation and assault to the cilia of sinus surgery. "This is a slowly progressive disease that does not get resolved quickly. It can seldom be reduced to a few simple elements."[2]

Ponikau shrugs. "You know, 500 years ago, everyone thought that the sun revolved around the earth. How could it all be wrong? People saw the sun going up and down every day. The paradigm shifted when Copernicus saw it another way. I'm hardly saying our findings are on par with Copernicus's—but look what happened! He was terrified to publish his theories.

After his death Galileo was imprisoned for embracing them. This is the way it is, in science."

Meanwhile, at the 2001 meeting of the American Academy of Asthma, Allergy, and Immunology in New Orleans, three Mayo immunologists, S-H Shin, Gerald Gleich, and Hirohito Kita discussed a study in which they found that the immune systems of patients with chronic sinusitis reacted to fungi by producing the same kind of eosinophilic inflammation Ponikau and Gaffey had observed. In healthy people, this reaction was absent, even when fungi were present. The study led to a $2.5 million grant from the National Institutes of Health to further investigate fungi's role in sinus disease.

Some speculate that the American diet may supply some answers in the rise of sinus disease and asthma, which are often linked. Massive amounts of antibiotics in the food supply—American cows, chickens, and pigs consume twenty-five tons of antibiotics a year—may be behind the increase. Trace amounts of antibiotic in meat, the theory goes, may kill the bacteria that prevent fungal overgrowth. To test the idea, doctors at Mayo hope to enroll the Amish of southern Minnesota in a study soon. The Amish breathe the same air as the rest of the population, but they grow their own food and raise their own livestock, shunning such "advances" as hormones and antibiotics. According to Ponikau, asthma and sinus disease are virtually nonexistent among older Amish, who maintain traditional ways. For the first time, though, asthma and sinus specialists in Rochester are seeing younger Amish, who have abandoned some conservative customs, as patients. Doctors caution that this study—and findings from it—are a long way off.

In the meantime, Sherris and Kern had started using topical antifungal sprays to kill the fungi, with promising results. In a study of fifty-one patients, all of whom had had repeated surgery, thirty-eight, or 75 percent, reported feeling better and needing fewer supplementary medications, such as decongestants and steroid nasal sprays. Even the doctors were surprised by how much their patients had improved. People who had had five, ten, fifteen surgeries were for the first time walking in for checkups without sniffling or wheezing.

Joanne Meyer, a sixty-eight-year-old piano teacher and mother of three, has struggled with the sinus trouble her whole life. "I lived on antibiotics," she says. "Flat out lived on them." She has been hospitalized for sinusitis, which invariably led to pneumonia (and the threat of meningitis) at least six times. During a three-week stay in an Indianapolis intensive care unit in the 1980s, she asked her doctor what the prognosis was. " 'You'll keep having these infections until one day you die,' " he replied.

In 1996, Meyer went to the Mayo Clinic. As a patient of Kern's, she was scheduled for surgery at least three times in as many years. But by every appointment, her CT scans had cleared a bit, and Kern agreed to put the surgery off. Finally, after the eosinophil discovery in 1999, he discovered a common fungi, aspergillus, in Meyer's nose, and prescribed a spray that used Itraconazole, an older antifungal medication. He told her not to expect immediate results, and when her problems didn't clear up, she was deeply disappointed. The antifungal spray was yet another nostrum in her otherwise loaded drug cabinet: prednisone, asthma inhalers, antibiotics, and decongestants. After a few months, though, Meyer noticed that she felt better, and had more energy—and realized she was cutting down on her other medications. "It's been fifteen months now without a single bad bout of asthma," she says. "For someone like me, that's a marvel."

Now divorced, Meyer says her illness put strains on her marriage. Her husband, she says, couldn't understand or cope with her condition. "I come from hardworking people who don't complain—that's my heritage—but for years all I could do was make it from the bed to the couch to the piano for lessons." When her children were small, she was sick constantly—so much so that she had to hire help in an era when only mothers took care of their children. "I got so depressed. I didn't wind up in the psychiatric ward or anything but you sometimes had to wonder about yourself. Why was I sick all the time? Why did it take me three months on antibiotics to get over a simple cold?"

Carol Van Camp, an airline systems programmer from Kansas, suffered from sinus, ear, and lung infections for years. With three small children in day care, she blamed her illness on her constant exposure to viruses. But when, in the late 1990s, her illnesses became so severe she was routinely hospitalized, she knew something was wrong. "I could never get well, ever," she says. "I'm not a complainer, but this drove me crazy." Her ears hurt so badly she couldn't sleep at night, and during her bouts of insomnia she wondered if she was losing her mind. Eventually, doctors prescribed tranquilizers. Still, she refused to give in to the disease. A lifelong athlete, she continued to break in her horses, or go on her daily four-mile runs. But she loathed the medication, particularly the prednisone. (Irritability is a common side effect.) "If people would cough near my desk at work I'd tell them to get the hell away from me. People thought

I was nuts. 'You don't understand!' I'd say. 'Your little cold will turn into a sinus-ear-lung ordeal for me,' I'd tell them. They thought I was some kind of hypochondriac."

After one sinus surgery, multiple procedures to insert ear tubes, and a move from an old farmhouse she loved to a newer structure where she might "feel better," Van Camp sought help at Mayo. Two years ago, Kern found a colony of fungi in her Eustachian tubes—they had traveled from her nose to the tiny crevice of the ear, causing chronic pain and inflammation. She, too, takes the antifungal medication. "It'll sound like some infomercial for a diet product, but I swear, this has given me my life back," she says. At first she was disgusted by having to squirt medicine into her nose every morning and night. Now, though, it's like brushing her teeth. Van Camp has not been on antibiotics for a year. "The one little cold I got went away *by itself,*" she says incredulously. "For me, that's just amazing. That never happened to me, not in fifteen years."

Now a double-blind randomized placebo trial is under way, but anecdotal results are promising.[3] "Folks come in clutching a newspaper article with the findings, having driven three days from Texas," says Sherris. He cautions that it is too early to declare victory. But in an ongoing study of Mayo patients in which disease ranges from mild to severe, mildly ill patients have responded as well as those who have the most difficult cases. Still, many doctors continue to scoff at the new data, pointing out that it has not been proven to be effective.

Stammberger maintains a scholarly resolve along with his cautious optimism. "How this treatment, after it is out of trials, will affect the millions with the disease, we don't know. So far, we have good news. But is it for everyone? We can't say."

Buzina, the microbiologist, makes some dry observations. ("I haven't any patients, only specimens. So I say what I like.") In Europe, medicine is funded by the state. Doctors get paid whether they perform surgery or not. "The situation is very different," Buzina says. "Even so, those who discredit the results will tell you that they object to the science. They are not going to say that they fear for their income." After so many years of training, many surgeons find it hard to believe that topical antifungal agents, applied daily, can alleviate their patient's symptoms. "Look," says Kern, pushing up his surgical glasses as we cross a Rochester street one bitter March day, "the closest thing we can liken this to is the news that *H. pylori*—a simple bacterium—caused ulcers, and could be treated with antibiotics. Ulcers weren't 'in your head' or 'from too much stress' or because of 'bad diet' that was somehow all your fault. They came from a bacterium in the gut. Nobody wanted to hear that news, either," he says, "especially all the heartburn drug makers."

The threat that the discovery will alter the surgical landscape may be more perceived than real. In fact, Kern and Sherris are still in the operating room two to three days a week. Sherris says that the team's revision rate appears to be less frequent than it used to be, but that it is too early to tell. "It's a new way of treating the disease. It's hopeful. But we don't pretend that it's a panacea."

In the continuum of medicine, Stammberger cautions, this is just another new discovery. Doctors don't know when the fungi get there, or when they start to cause problems. In Graz, doctors are even starting to look in the noses of newborns to see if babies are born with fungi they got from their mother's bodies.

"There is so much we don't know," he says. "But as we reflect on medical history and the treatment of this disease, you have to wonder." Like all science of the nose, surgery of the sinuses has undergone its own recent evolution. Stammberger believes that after some initial missteps, the field is now on the right course, with the early, overly intrusive surgical methods giving way to treatments that combine antifungal medicines and antibiotics with more conservative surgery.

"We know now that the old approaches tended to make things for the sick person even worse. Chopping out the innocent turbinates, stripping the membrane—none of this was done with malice. But we kind of had to go through all that in order to get to this point. Discoveries like this put things in a new perspective."

Even those who fully embrace the fungal theory insist that it doesn't mean surgery is obsolete. Polyps must still be removed; deadened, chronically infected tissue must also be excised before sick sinuses can hope to heal. In order for the antifungal treatment to be effective, it has to be able to reach areas that are often covered with scar tissue. "You wouldn't use a topical medication on skin that was coated with a layer of Jell-O, would you?" Kern asks. "Sometimes, there is no way around surgery."

Stammberger agrees. "But certainly today you hope that there is a brain behind the hand that is cutting you," he says.

"As we look back on the history of treating this disease, you come to the painful conclusion that the twentieth century was a century of butchery." Stammberger gives a sigh. "As someone who helped to improve these techniques, I hope I can eventually be someone who helps make them less necessary."

Part Three

The Commercial Nose

nine

Smells Like Money

Whether we like it or not, our noses today are working overtime. The early nose functioned to sniff out danger, but today's nose is bombarded with scents that are supposed to help seduce, relax, refresh—practically everything short of unload the dishwasher. Botanical candles at Pottery Barn vow to transform rooms. Fragrance-doused paper pine trees for the car's rearview mirror are stacked next to Doritos at the gas station. The Gap displays eau de toilette and body lotion called Heaven and Dream right next to the checkout line. From upscale boutiques to Wal-Mart, you can buy an aroma to help set mood or suggest a place (it is not clear who did the research on heaven). But this burst of scent is a recent phenomenon in America. In less than fifty years, fragrance journeyed from crystal bottles behind the glass counter at Neiman Marcus to the street.

The American attitude toward such luxury got off to a slow start. But each epoch and cultural turn powerfully influenced America's relationship to fragrance, from the first transatlantic flight to postmodern sexuality, from dazzling materialism to pared-down simplicity.

Annette Green has seen it all. While many industries see trends and leaders come and go, Green has endured as the scent trade's evangelist for more than forty years. In 1961, Green took charge of the Fragrance Foundation, a nonprofit organization designed to educate the public about the importance of scent, and has served as the industry's chief spokeswoman, historian, and archivist ever since. (She has even established a fragrance museum at her office in downtown Manhattan.) Her mission in life, as she sees it, has been to liberate scent from its imperial perch in fancy department stores to envelop the masses. She has quietly guided the way Americans buy, and use, scent: annual retail sales in the personal fragrance industry—perfume, eau de toilette, and soap—have gone from $250,000 in 1950 to $6 billion in 2000.

Walking into the Fragrance Foundation is like walking into a spectacular restaurant. A pale green carpet hushes noises. Giant backlit bottles of amber perfume glow against the cool white walls of the two-story office in Manhattan's Murray Hill. A scent is pumped into the office daily, so delicate it is all but unnoticeable. Is that the tiniest bit of spilled perfume—those bottles really are everywhere—or your imagination? Green is mum on her own favorite scent, and no one else in the office will say what the fragrance is. The secrecy emits its own mystique. "It's all Annette," says Theresa Molnar, the executive administrator for the Sense of Smell Institute and Green's right arm. "She thinks of everything."

Green, in her eighth decade, has amazing energy. She is at her desk before eight, and often stays past seven, long after the staff has left. A trim, small woman with short gray curls and impeccable makeup, she wears Ferragamos, knots Hermès scarves around her neck, and sports the hands and ankles of a woman half her age. Her nails are painted a pale pink, her fingers a flash of gold rings.

Where others divulge nothing—industry insiders are as covert as CIA operatives—Green is forthright and bold, and speaks her mind on trends.

Green's affinity for fragrance began at birth: her mother went into labor as she stood sampling perfumes at a Philadelphia department store. Though luxury was off-limits for most women in the midst of the Great Depression, Green's mother, a milliner, saw perfume as essential. When Green was small, she knew the best way to make her mother happy was to save for a special bottle of the stuff, for Christmas or her birthday. "She just lit up when she'd open it," Green says.

Her mother was the exception. During the first half of the twentieth century, American interest in scent was minimal. Most products were French, and by association so blatantly erotic that they were anathema to rugged American sensibilities. While the French had been using perfume for four hundred years, the Americans had a decidedly different tradition. (As recently as the 1800s, some descendants of American Puritans still viewed body odor—man as God made him—as an outward sign of godliness.)

In 1949, four French perfumers, Caron, Chanel, Coty, and

Guerlain, established a small consortium, which they named the Fragrance Foundation. Its first director traveled the country on the ladies' luncheon circuit, speaking on the powers of scent. But the notion of wearing Shalimar or Joy (launched in 1935 as the "most expensive perfume in the world") seemed an unspeakable indulgence, especially in a period when women were returning from the wartime workplace to the hearth. And as the country set out to Americanize recent urban immigrants, the goal was to get people *not* to smell. Foreign language newspapers from Manhattan to Milwaukee urged new Americans to wash with products that conveniently doubled as both soaps and household cleaners.[1]

The Fragrance Foundation struggled to promote its wares in the United States, but bridging the cultural gap was neither easy nor swift. In the postwar period, scent was a by-product of the ultimate ideal: a squeaky clean image both at home and in person. Perfumed soaps like Tide and Cashmere Bouquet had been around for decades, but after the war chemical companies devised a slew of new cleaning products for their new market of college-educated housewives. In *The Feminine Mystique,* Betty Friedan documents the advertising of new sprays and powders that would allegedly clean better than old-fashioned elbow grease and vinegar. By buying these new cleansers, scented with synthetic pine and lemon, women were made to feel as if they were doing themselves—and their families—a great service. Housewives could put their brains to use, right in their own home laboratories.[2]

Fragrance on the body was another matter entirely. Returning GIs had brought home fancy bottles of French perfume after the liberation of Europe, but for the most part, even years after the war they remained unopened on dressing tables, too intimidating to use.

Meanwhile, Green, an aspiring journalist, had moved to Manhattan from New Jersey, where she hoped to land a job at a fashion magazine. Such work was highly sought after—there were few respectable jobs for young ladies, and editorial assistantships at *Vogue* and *Harper's Bazaar* were chief among them. Instead of waiting for an opening, in the early 1950s Green took a job at a publication called *The American Druggist,* where she soon wrote her own column. One day, one of her bosses asked Green to help dress windows of a New Jersey drugstore. She agreed, and once the task was completed, she turned to studying, and chatting with, the teenagers who passed time at the store's soda fountain. As it turned out, when girls had finished with their flirting and floats, they turned their attention to cosmetics. In those days, such products were kept behind the counter, much as condoms and pregnancy tests often are today. In her early twenties, Green watched the girls with fascination as they giggled and tried on lipstick testers. But only a brave few mustered the courage even to ask for anything, let alone buy an item.

After a few trips to the store, Green suggested to the owner that she might step in as a salesclerk herself. She thought her presence as a young woman would make the girls feel more comfortable—and be more profitable for the store. Her hunch was right, and soon she was a fixture on Saturdays, laughing and joking with the teenagers as she racked up sales of compacts and rouge. Green wrote about the simple logic in her column, deftly suggesting that cosmetics would yield even more sales if they were moved onto open shelves. Girls with crushes on good-looking soda jerks were bound to visit drugstores. If they could legitimize their "stopping in" with a purchase, all the better.

Green earned a reputation as a thoughtful marketer. But writing was her real love—or so she thought—and soon she landed jobs with publications elsewhere. Within a few years, however, a French perfumer, Lenthéric, offered her a job in public relations. She took it. "They paid me a heck of a lot more money than I'd ever made as a journalist," she says. Eventually, she started her own public relations firm, and in 1961 took on the Fragrance Foundation as a pro bono client and soon began serving as its executive director. Green knew that perfume, like cosmetics behind the counter in the drugstore, was daunting to most American women. Yet French women, whether they were doctors or housewives, glided through Paris swathed in scent. "It's as natural to them as breathing," she says. For Green, fragrance was magical. She hoped that one day, Americans could see it the same way. As it was, women only used a tiny drop behind the ears before going out—and then only for special occasions.

But as the country redefined itself in an era of middle-class comfort, women strove to keep up with new pleasures, from cars to TVs. Magazines were more than happy to dispense advice about how to spend money on advertisers' products. Perfumers, of course, paid for many ads, and *Glamour* magazine encouraged readers to be daring with the stuff. A 1957, article advised: "The most important rule in the art of applying fragrance is to apply it. Don't simply admire the bottles on your dressing table. And learn that the only thrifty way to wear perfume is to use enough to count—a touch behind the ears is a lot too little and a waste of this valuable, functional invention."[3]

A Shift

Just as Green was wrestling with how best to organize the foundation, a new development transformed America: airplane travel. Pan American World Airways had recently completed its first transatlantic flight from New York to Paris, cutting jet travel time in half and creating a huge upsurge in Americans traveling abroad. Films helped deepen the appeal—from *Sabrina* and *Roman Holiday* to *Gigi*. Paris beckoned above all. American women returned from visits with flacons of perfume that they had sampled and chosen themselves. For the first time, they began to wear it regularly. Chanel No. 5, the French sensation (introduced in 1921, it was the first scent to contain aldehyde, a synthetic molecule), enjoyed instant fame on the other side of the Atlantic after Marilyn Monroe told a reporter that it was all she wore to bed. European manufacturers began to recognize the American market, and by the end of the 1950s they had introduced thirty new women's fragrances. In the United States, Estée Lauder launched Youth Dew, a heavy floral scent purportedly worn by Joan Crawford and Gloria Swanson. Avon (which began in the 1880s as the California Perfume Company) became a trusted name by selling fragrance and makeup to housewives directly, with no intimidating salesclerks. Saleswomen were peers who went door-to-door proffering little product samples, so women could see which colors and scents their husbands liked best.

The packaging of an era's fragrances mirrored its architecture, striving to be both modern and "functional." Even cosmetics companies entered the fragrance market, designing gold or silver-toned lockets that held a scented wax that could be smeared across the skin. They were inexpensive, and fairly disgusting, but few women got past Mother's Day without receiving one as a gift.

Green knew that with money to be made, and glamour to be had, Americans would start wearing real perfume sooner or later. Using her training as a journalist and her skills as a marketer, Green spoke to society columnists at major newspapers everywhere. Fragrance, she told them again and again, was as important in a woman's wardrobe as the proper shoes.

By the early 1960s, companies had even introduced scents for men. Far from Dad's stodgy Old Spice or medicinal Aqua Velva, these were uncorked machismo: Aramis, named for the Third Musketeer, was supposed to evoke the manliness of a bygone era. English Leather, though available at the drugstore, appealed to the proto–Ralph Lauren suburban patrician. And Brut's celebrity endorsers said it all. In commercials, Joe Namath and Wilt Chamberlain, totems of American masculinity, slapped Brut liberally on their cheeks as women clustered nearby. Hai Karate, introduced in 1968, at least had a sense of humor. It used "Laugh-In"-like ads to show that men who wore it were so attractive they had to resort to martial arts kicks to fend women off.

But as art mirrors life, industry reflects culture. As soon as perfumers had established a market, Woodstock and antiwar marches jolted American society. At campuses and protest meetings far outside tidy suburbia, perfume was the last thing on people's minds. Fragrance, however, lingered: its ability to transform image and mood somehow stuck with consumers, no matter how antiestablishment they were. Soon jasmine candles, sandalwood massage oils, and patchouli incense sticks mingled with marijuana as an accessory to the sexual revolution. In the early 1970s, perfumers launched breakthrough scents like Cachet, a woodsy fragrance said to "react" with a person's individual chemistry. (In fact, all scents vary from person to person, reacting differently to each wearer's skin.) Nevertheless, Cachet advertised that it was "as individual as you are."

Scent was now accessible to everyone—in price, location, and packaging. While early French perfumes bespoke elegance and sophistication—Baccarat and Lalique often designed the bottles—vessels were no longer demure and feminine. Now, they were simple, modern, utilitarian, and often downright phallic. Many were cylindrical and topped with large half-spheres of plastic or metal. They could be purchased everywhere—at drugstores, discount outlets, even supermarkets. If the scents of yesteryear had appealed to the inner glamour of a trapped housewife, the fragrances of the seventies appealed to Everywoman, Everygirl, and even Everyman—and their inner desires.

Revlon introduced two fragrances, Charlie and Enjoli, with blatant appeal to emerging feminist sentiments. With its perky pantsuit-clad model, Charlie ads invoked the independence of the new young woman. In one commercial the blonde "Charlie Girl" carries a briefcase, and strides happily along a city street. The spot ends with her patting a male coworker on the behind as she turns to grin at the camera. Enjoli ads had a similar message: You *could* have it all. Its sassy blonde model, in a navy blue suit and pumps, comes home and begins slinging pots around her kitchen while she sings Peggy Lee's classic: "I can bring home the bacon / fry it up in a pan / and never let you forget you're a man." By the parting shot, she has changed into a slinky cocktail dress, and thrusts a bottle of Enjoli toward her viewers.

Suddenly, scent was everywhere, even in snacks and makeup. New candy, such as Starbursts and Tic-Tacs, were loaded with artificial flavorings and potent fruity "aromas." Fragrance permeated toiletries like shampoo; "Gee, Your Hair Smells Terrific" and Herbal Essence were top sellers. Flavored lip glosses borrowed scents from popular culture: Dr. Pepper and Bubble Gum were two favorites with adolescents. Some colognes were specially marketed to teenage girls. No sixteen-year-old's dressing table was complete without a bottle of Love's Lemon Fresh or Baby Soft, a sweet, treacly scent, colored pink, that theoretically bottled innocence. Its ads sent a somehow different message: a young girl, dressed in a low-cut white dress, sat lasciviously licking a lollipop. The box read: "Because innocence is sexier than you think."

But nothing changed the landscape of fragrance like musk. Medieval Arab traders first sold musk as an aphrodisiac. (The word "musk" is derived from an ancient Indian word for testicles; the deer's gland is located right next to them.) Two entrepreneurs, Bernard Mitchell and Barry Shipp, took note of hippies buying vials of the stuff (or rather, its synthetic substitute) from incense and head shops in Greenwich Village, and introduced Jovan Musk in 1972. Rather than use clean-cut blond models, Jovan ads, which ran in magazines from *Playboy* to *Redbook*, left nothing to the imagination. Neither did the bottles. The company introduced his-and-hers gift sets with flasks that nestled suggestively together. One said that it was the "provocative scent that instinctively calms and yet arouses your basic animal desires. . . ." Another promised: "It may not put more women into your life but it'll put more life into your women." And a commercial opened with the question, "What is sexy?" It showed a man ogling a woman in a skimpy dress as she washes her car, the hose squirting suggestively.

Fragrance in the 1970s reflected an open sexuality, but also another feature of the decade: frugality. At a time when people struggled in gas lines and wore buttons vowing to "Whip Inflation Now," the industry was full of affordable colognes and eaux de toilette. Now that more people could afford it, attitudes toward scent had changed dramatically—enough so that Green inaugurated an annual awards ceremony to honor the year's most innovative fragrances. Modeled after the Oscars, she called it the Fifi Awards.

The 1980s ushered in an era of conspicuous consumption. Designer products expanded from clothes to sheets to eyeglasses, and inevitably, to scent. It had taken twenty years, but Green finally had her wish: American women had begun to think of perfume as vital. If you couldn't afford Calvin Klein or Oscar de la Renta couture, you could at least wear status on your wrists and neck—and almost everyone did.

Scents were just as dramatic as the big hair and tight jeans that were in style. Bottles had extravagant designs and bright colors. Even their names evoked power: Poison, Obsession, and Decadence. "It was a voluptuous, egocentric time and fragrances were too," Green says. "People wanted you to be able to smell them from across the room."

One potent scent, Giorgio, achieved just that. Available only by mail order or in trendy boutiques, it had a snob appeal akin to that of clubs modeled after Studio 54. Shortly after its introduction in 1982, Giorgio founders Fred and Gale Hayman gambled on a new technology they believed would popularize their scent like no fragrance counter ever could. With $300,000 of their own money, they invested in a "scent strip" researchers said could deliver a whiff of the fragrance straight to consumers' nostrils—millions of them. Unlike the crude "Scratch 'n' Sniff" technology of the 1970s, the perfumed strips came on double-thick perforated pages that could be bound into glossy women's magazines like *Vogue* and *Glamour.* When torn away and rubbed across the wrist, they gave consumers an actual sample of the fragrance.

The gamble paid off. Within Giorgio's first three years, the fragrance generated $80 million in sales. Soon the heavy scent was so ubiquitous that restaurants in New York and Los Angeles posted signs: "No Smoking. No Giorgio."

For the first time, Green says, masses of people had the opportunity to sample fragrance away from the retail marketplace. Men could try it out in the waiting room of their doctor's office instead of at a counter they were too embarrassed to approach. Within a few short years, every perfumer distributed the scented strips. By the end of the 1980s, you could try seven or eight of them, all for the price of a magazine. For a time, the novelty even boosted magazine sales.

As Green reveled in the success of fragrance in America, she also became aware of the small groups of scientists studying olfaction and behavior. If perfume could make people feel good about themselves, surely scent used on a larger scale had implications for moods as well. Though olfactory science was in its infancy, Green met with researchers at Yale, the University of Pennsylvania, and Georgetown to discuss how the foundation might reinforce the researchers' efforts. At a meeting with the foundation's board of directors, Green asked the industry to support the new discipline.

"Aromatherapy" had been in use in Europe for many years but had only recently come to the United States. Green never liked that phrase, thinking it connoted folklore rather than science. She coined "aromachology" to describe the work she hoped the foundation could support with a new research arm, the Olfactory Research Fund (since 2001, it has been called the Sense of Smell Institute). Green envisioned offering grants to scientists of olfaction, particularly those whose research bolstered the positive effects of scent.

The fund has since sponsored hundreds of studies on smell and behavior, ranging from the role of fragrance in women's sexual fantasies to the sense of smell in space.[4] New York's Memorial Sloan-Kettering Cancer Center found in 1994 that the smell of heliotrope, a vanilla-like scent, significantly reduced the anxiety of patients about to undergo an MRI. Now technicians pump the aroma into the claustrophobia-inducing chamber before loading patients into it. By 2001, the fund had financed 51 projects for $1.3 million.

Indeed, the MRI study helped launch a whole new industry within an industry: helping overworked Americans unwind. By 1997, several companies had introduced a wide range of aromatherapy products. Coty's Vanilla Fields claimed to calm, and Shiseido's Relaxing, a mix of bamboo, tea rose, and cucumber, promised to soothe away stress. Estée Lauder's sporty division, Clinique, launched Happy, and Origins, its upscale bohemian line, released aromatherapy products (including aromatic gumballs) in 1998.

Home-furnishing stores started selling scented candles and room sprays—updated Renuzit. Aromatherapy products—creams, tinctures, shampoos, and lotions—soon crowded the toiletries aisles at drugstores and supermarkets. Smell, suddenly, had overwhelmed the marketplace. An industry trade publication found that home fragrances such as candles, incense, and bathroom soaps increased more than 20 percent in 1998. A Connecticut candle and potpourri manufacturer, Blyth Industries, reported sales of nearly $1.2 billion in 2000. And NPD, a marketing research firm, estimated that aromatherapy products within the beauty industry grew 8 percent between 1999 and 2000.

Green loves the barrage of fragrance, though she sniffs that "a lot of companies have jumped on the aroma bandwagon." New research, she then adds, has simply helped manufacturers to make more and better products.

One would hope. By the mid-1990s, celebrities such as Sophia Loren, Elizabeth Taylor, Cher, and Mikhail Baryshnikov had each put out their own fragrances, but few of these products took off. Cher's Uninhibited seemed anything but. Its claim was offset by its bland scent and detailed Art Deco bottle. Baryshnikov was celebrated for his soaring flights, both artistic and political, but his perfume, Misha, fell flat. Meanwhile, Elizabeth Taylor's fragrances, Passion, followed by White Diamonds, were hugely successful. People wanted to smell like the beautiful icon from *Cleopatra,* but not necessarily like Cher.

Scents in the 1990s tried to evoke simple pleasures. Crabtree & Evelyn Ltd. has a line of "kitchen scents," ranging from Patisserie to Salad Greens. The marketing pitch is domestic rather than seductive. Ads say: "We all love those delicious aromas wafting from the kitchen. Now you can enjoy them anytime—even when the cook's off duty." The reality? Salad Greens smells more like Seven Seas bottled dressing than arugula dressed with balsamic vinegar and olive oil. And a stroll through Sephora, the ultramodern French cosmetics chain with branches worldwide, is almost too much for the nose to bear. There is Fresh Rice Oil ("to unblock 'chi,' "), Brown Sugar Body Polish ("rejuvenating and relaxing"), and Chocolate-Orange-Chocolate bath foam ("bewitching"). But Green approves. "If you're dieting, you want rich aromas," she says. "Sometimes you just want the smell or taste of something you love—not necessarily to eat it." Indeed, a recent institute study conducted by Susan Schiffman, a professor of psychology who specializes in eating disorders at Duke University Medical School, found that inhaling scents can be nearly as satisfying as eating.

The trend doesn't end there. Demeter, a New York company, comes out with new colognes every season. Although they are called "Pick-Me-Up" sprays, some, it seems, are intended more to lift spirits with their wit than with actual scent. Demeter's colognes range from Dirt to Popcorn and Altoids, "the curiously strong mint." (The company even has a collection they call their "Attitude Adjustment" line. Touted as "anti-aromatherapy," the scents range from Gin-and-Tonic for the recently jilted to Brownie for the "Never Happy.")

And in recent years, the notion of scented products has drifted into household cleaning products. Tired of your wooden floors smelling like Murphy's Oil Soap, and your bathroom like Fantastick? The makers of such products as Palmolive and Lysol have launched new scent lines, although it's anyone's guess what "Orchard Fresh" dish detergent and "Spring Waterfall" countertop cleanser are actually meant to evoke (or are going to make you forget you had to clean up after dinner).

Upscale retailers have also introduced new products. Elegantly packaged cleansers—everything from scented water to put in the steam iron to window washer—are sold at high-end stores like Williams-Sonoma and Anthropologie. They smell great, and are, of course, expensive: Williams-Sonoma's special basil and verbena countertop cleaner costs $9.50 a bottle; their chamomile ironing water costs $12. Consumers at Nordstrom's cosmetics counters can now pick up such items as Caldrea's Green Tea Patchouli dish soap ($8), and Jane French Laundry Wash ($16), along with their mascara and lipstick.

Although these products represent a small share of the market, Green expects the niche to grow. She laughs at the innovations. "The nose is individual. Why not pamper it? We now know that the nose can improve our whole quality of life. If popcorn's what turns you on, hey, why not?"

At the beginning of the twenty-first century, Green thinks that people today have reawakened to the significance of their sensory selves, in cuisine, in architecture—and in their noses. She feels it an inevitable humanizing response to the rise of the computer. "We still have primal urges," Green says of the Internet, "and one of the most pronounced is smell." Perhaps both worlds will meld someday—a French cinema company, CanalPlus, is exploring the possibility of a box to deliver scents that correspond to the audiovisuals. Like Smell-o-Vision, the 1950s precursor of the idea, moviegoers may soon be able to see a chef chop and sauté onions on-screen—and smell it as well. (A Silicon Valley firm, DigiScents, enjoyed brief fame in the late 1990s when it introduced a computer device containing a mixture of chemicals that allowed a person to smell certain odors while browsing the Web. It went bankrupt after two years.)

Cyrano, a company in Pasadena, California, is developing a computer chip that can smell. A simple version of the device already exists in smoke alarms, but Cyrano hopes to eliminate the human variations—colds, fluctuating hormones—that can influence a person's ability to sniff out items gone bad, from food and wine to beauty products. Lancôme, the French cosmetics company, and Starbucks, the Seattle coffee giant, have both bought models of what Cyrano calls its "e-nose." The device is capable of distinguishing only a fraction of the 400,000 smells humans can distinguish, but researchers believe that computer advances will yield better results soon.

And then there are the increasing commercial applications for pheromones. Fragrance companies are adding synthetic pheromones to perfumes and colognes. David Berliner, the researcher and venture capitalist in Menlo Park, started a company called Erox; its products include a line of pheromone-laced scents called Realm. A British company called Kiotech now markets a product it calls Sex Wipes, a baby wipe–like tissue soaked with pheromones. It sells them in rest rooms at nightclubs and bars—right next to the tampon and condom machines. Winifred Cutler, the biologist and behavioral endocrinologist, has founded Athena Institute, which manufactures fragrances with chemically reproduced human pheromones. Its men's line, called Athena Pheromone 10X, promises, like the swinging Jovan ads in *Playboy* of the 1970s, to "Raise the octane of your aftershave." The women's version, Athena Pheromone 10:13, conveys more of a *Ladies' Home Journal* sentiment: "Let human pheromone power enhance your sex-appeal and increase the romance in your life."

No scientific studies exist to show whether the products "work" or not. One day, though, on a crowded commuter train to Manhattan, I sat next to an athletic-looking man as I sifted through some material on pheromones, including Realm. My seat partner politely asked if I was a scientist, so I told him I was researching a book. The man, Ted Thomas, turned out to be a dancer with the Paul Taylor Dance Company in New York. He said that he used Realm regularly, and that when he did, women approached him freely to strike up conversations, telling him deeply personal details within minutes of meeting—medications they were on, family histories, job woes. "And they always finish with, 'I don't know why I'm talking so much. I don't even know you,' " he said. To be sure, Thomas is a handsome man—almost intimidatingly so—and, as a professional dancer, he carries himself with uncommon grace. I remarked that he very likely had little trouble attracting women in the first place. Which came first? I asked. He blushed and said, "Well, *more* women talk to me now." For now, there is no definitive answer.

Berliner welcomes such anecdotes. "I am not at all surprised," he says, not at all modestly. He thinks for a moment, then adds: "But Realm is not an aphrodisiac. The same substances which might attract one person may repel another." Still, Realm, which is sold only in department stores and over the Internet, made Erox $20 million in 1996, only a year after its debut. (Berliner won't divulge current company revenues, but a recent *Wall Street Journal* article put the figure at $40 million.)

But from pheromones to Demeter's kitschy Funeral Home scent, there may be some signs that the industry has gone too far. Millions of Americans are believed to suffer from MCS, multiple chemical sensitivity. They experience myriad symptoms, primarily inflammation of the nose, sinuses, and lungs, when they come into contact with low levels of chemicals. Estimates vary, but government statistics show that anywhere from 2 to 6 percent of the population might be affected. A whiff of anything scented, from cigarette smoke to laundry detergent to perfume, can make those with MCS ill.

Now, companies that once spent millions on coming up with the most appealing fragrance for their products make "scent-free" soaps, deodorants, and detergents. Restaurants, doctors' offices, and some corporations now prohibit their employees from wearing scented products. In the heyday of scent strips, so many mail carriers—and subscribers—complained of allergies and headaches that magazines began to limit them. (*The New Yorker* even had a scent strip hotline, for subscribers who wanted odor-free issues.) Some towns are considering banishing artificial scents altogether (Halifax, Nova Scotia, already has).

This backlash against the perfume excesses of the 1980s and the aromatherapy bombardment of the 1990s may be a sign of things to come. Like a laminated menu with dishes numbering in the hundreds, some say there are too many scents—the nose needs relief. Sales of perfume have begun to show signs of sluggishness. While it could reflect a downturn in the economy—clothes, jewelry, and air travel have slowed too—some critics, particularly the scent-free-space advocates, say that people have had their fill of fragrance.

Ironically, some new research points to evidence that sexual attraction—the promise of many perfumes—is actually based on natural odors. Synthetic ones, therefore, might hurt rather than help your chances of attracting the right mate. Isn't the romance and mystery surrounding the fragrance industry, then, just a clever ruse? Might wearing colognes and perfume be actually counterproductive?

None of this seems to worry Green. She has seen many things come and go, and she retains a sunny confidence about the future of perfume. Fragrance, she says, is here to stay. "Look at the changes in the industry in the past forty years," she says. "We've gone from women afraid to wear perfume to now having scented body lotion at the lingerie shop. When you go to pick up a couch, you can pick up potpourri for your car." Green shrugs. "I get a lot of energy from my work—the sense of smell itself. You just have to keep thinking, 'What will people want and need tomorrow?' "

That, of course, is anyone's guess. If aromas can affect sexual fantasies, why not research odors that could minimize aggression, or even combat fear? Scent has held its allure for millennia. "The world moves on," Green says, waving a perfumed hand in the air. Every epoch has its successes, from Nero's rose petals to Giorgio. The future, she says, will come from science, from learning how fragrance affects our moods and our behavior.

Ads and bottles help, but in the end, people will respond to how scents make them feel, dream about, and hope for—or what it makes them remember. Perhaps Proust gave the best explanation for how an industry founded upon the esoteric can flourish. Perfume, he wrote, "is the last and best reserve of the past, the one which when all our tears have run dry, can make us cry again."

ten

The War on Stink: Banishing Body Odor

The flip side of fragrance, of course, is stink. Human beings are naturally pungent. Food particles affect the breath. Women menstruate. People sweat. The average adult body has between 3 and 4 million sweat glands, capable of producing four gallons of fluid a day.

But Americans, in general, loathe the smell of human beings. We do anything to keep from smelling like ourselves—bathing, scenting, and deodorizing more than at any other time in history. Even so, humans are still the most "highly scented of apes," Michael Stoddart writes in his seminal book on human odors, *The Scented Ape: The Biology and Culture of Human Odor.*

Each person has an odor as individual as the whorls on his fingertips. Body odors differ from culture to culture and even among race: they are influenced by genes, environment, diet, and bacterial flora. (While people generally blame perspiration for smelling bad, sweat itself is odorless. In fact, the fluid excreted by the eccrine, or sweat glands, is nearly 99 percent water, but it begins to stink when it breaks down the bacteria that rests on the skin's surface.) The culprit for your most pungent odor is the apocrine glands, which release not only perspiration but also fatty substances that provide the perfect environment for bacteria to multiply. While sweat glands begin working at birth, apocrine glands start functioning only at puberty. The apocrines are concentrated mainly in the armpits and around the genitals, and it's no coincidence that these areas usually sport tufts of wiry hair—hair that soaks up the thick, oily apocrine secretion and disperses it into the air from a vastly increased surface area. The placement and texture of the hair leaves endocrinologists with little doubt about the odor's role as a sexual attractant—at least millennia ago.

Caucasians and Africans have more aprocrine glands than do Asians, and consequently, some experts say, have sharper odors. In fact, having body odor in Japan was once thought to be a sign of poor genes and bad character, and immediately disqualified one from military service. Recently the Japanese concern about stinking prompted a Shiseido researcher to isolate the smell of growing old: in particular, nonenal, an unsaturated aldehyde. While nonenal is a traditional component of body odor, men over forty were found to emit particularly concentrated amounts. Shiseido and other fragrance companies were quick to capitalize on fears of smelling like *ojisan,* a term that literally means "uncle" but also implies a square old man who is hopelessly inept. New nonenal-neutralizing products quickly appeared on shelves of Japanese drugstores and department stores—including a line of antimicrobial men's underwear. Ads for the briefs featured a young woman surrounded by middle-aged men, arms hanging onto subway straps, in a crowded commuter car. Overwhelmed by their odor, she finally screams, wrests her Walkman off her head, plucks the foam ends of her ear pieces off, and stuffs them up her nose.[1]

Elsewhere, it is foreigners who smell the worst. The Chinese, for example, refer to Westerners as those who "stink like butter." Butyric acid in butter and other dairy foods makes Americans and Europeans reek to Asians who eat no milk products. Likewise, odors from garlic, onion, strong-smelling fish, and spices are carried through the bloodstream and excreted through the sweat glands to the pores. In the United States and in Europe, disdain for immigrants who eat such sharply flavored food has often expressed itself as disdain for their objectionable "smell."

Yet around the world, body odors are meaningful in ways many Westerners might find repulsive. In Burma, a common greeting is "Give me a smell" ("Come close enough so I can smell you"). Members of a New Guinea tribe bid good-bye to each other by placing their hands under each other's armpits and rubbing themselves with the other's scent. (By contrast, a German expression for "I can't stand the guy" is *ich kann ihn nicht riechen,* literally, "I can't smell him.")

Americans are often quite disturbed by the smells of others, which poses delicate personnel problems. In offices and factories around the country, managers struggle with the stink that dare not speak its name: the body odors of employees. Whether caused by a diet rich in spices, poor hygiene, or too much perfume, supervisors everywhere wrestle with this most sensitive of issues.

Few executives are even willing to discuss the problem openly for fear of embarrassing colleagues or employees. To inquire about the source of body odor can trigger civil rights liability—if the problem is caused by medical conditions or certain foods in the diet, it is covered under the Americans with Disabilities Act. One Portland, Oregon, manager suggests that employers must first verify if the problem exists—sometimes people are downright malicious. If it does, she says, "proceed with caution." And if the employee volunteers that his or her body odor is the result of a medical problem, she advises, "Shut up!"

The First Deodorants

The human journey from reeking stench to intolerance for odors was relatively quick. Prior to the nineteenth century, Westerners merely disguised their body odors with strong perfumes, if they bothered to do so at all. (René Laennec, a nineteenth-century cardiologist, found a way to ease the olfactory assault of diagnosing heart problems the old-fashioned way—putting an ear to the thorax—when he devised the modern stethoscope. It put a safe distance between the doctor's nose and the patient.) Baths were weekly, at a maximum, until indoor plumbing became commonplace in the 1920s and 1930s.

Until then, anyone concerned with such a mundane matter as body odor relied on household products—one popular way to fight odors was a mix of ammonia and water splashed onto the tender skin of the armpits. In 1888, a Philadelphia doctor came up with a commercial concoction, a thick, waxy cream made with zinc oxide, a microbial designed to kill odor-causing bacteria. He called it "Mum," and relied on his nurse to sell and spread word of the product. It not only caught on, it worked, and soon many were using Mum and other creams like it. Applied with the fingers, the first deodorants were cumbersome, and contained caustic chemicals that not only stung but ate through clothing.

By mid-century, other inventions, many from the war, led to the development of more practical applications. A pointed tip, loaded with a ball bearing, was able to deliver a controlled amount of liquid through a tube in the first ballpoint pen. (The British Air Force was one of the first wholesale consumers of the pens, once airmen who used them at high altitudes discovered that they didn't leak.) Proctor and Gamble used the mechanism to deliver the first roll-on deodorant, Ban, in 1952. It became a top seller. Now, though, deodorants had an added ingredient: chemicals designed to quell not just odor but sweating itself, by reducing the amount of perspiration that reaches the surface of the skin.

Americans, it turned out, were more than eager not to smell bad. Listerine, for example, produced by the Lambert Pharmaceutical Company, was first used in the 1870s as an antiseptic in hospitals and homes. In the 1920s, it was relaunched as a mouthwash. Ads showed a young woman gazing at her reflection in a mirror. "What secret is your mirror holding back?" it read. The text went on to say: "She was often a bridesmaid but never a bride. And the secret her mirror held back concerned a thing she least suspected—a thing people simply will not tell you to your face. That's the insidious thing about halitosis (unpleasant breath). You, yourself, rarely know you have it. And even your closest friends won't tell you." Paranoia suddenly boosted sales from $100,000 in 1920 to $4 million in 1927. Listerine's ingredients hadn't changed a bit, but its concept had.[2]

During the Great Depression anxiety intensified, and soap companies stepped up their campaigns as well. Lifebuoy soap coined the phrase "B.O.," for body odor, and echoed Listerine's hint with a 1931 ad that warned: "Don't risk your job by offending with B.O. Take no chances! When business is slack, employers become more critical. Sometimes very little may turn the scales against us."[3] A radio ad for Lifebuoy made B.O. sound terrifying. It used a foghorn with a screechy voiceover shouting "Beeee-ohhhh!" The soap became the country's most popular bath soap.

In prosperous postwar America, worries about job security had long since faded. The technological advances that propelled the economy also revolutionized the war on body odor. The new antiperspirants, which contained aluminum salts, were thick and slow to dry, and customers complained that they stained clothes. But a second wartime discovery added another alternative to the list—the delivery of a product through aerosol spray. Though it had been in use since the late nineteenth century, in 1941 inventors Lyle Goodhue and William Sullivan patented the aerosol, which was used to deploy insecticides for U.S. troops at risk for malaria during World War II. By the late 1950s, manufacturers had deployed antiperspirants in cans as well, and by the early 1970s, they accounted for 80 percent of

Depression-era Lifebuoy ads like this one did their best to highlight the social ramification of body odors. The threat of spinsterhood was at the top of the list.

the market. Heightened awareness of air pollution in the mid-seventies led to a ban on spray cans using aluminum zirconium, a main ingredient in the antiperspirants. Today, some 95 percent of adult Americans use some form of antiperspirant or deodorant, according to the Gillette Company.

Americans go to great lengths to stop sweating—and stinking. Some have a condition called hyperhidrosis, or excessive sweating, and for extreme cases doctors sometimes remove sweat glands or the nerves that stimulate them. The procedure has fallen out of favor since the introduction of a prescription product called Drysol, which, when applied correctly, can reduce sweating for up to a week.

Even cosmetic surgeons have discovered a slice of the sweat-free market: the toxin botulinum, or botox. First they injected it in wrinkles to paralyze the muscles causing them; then they realized the same treatment could relieve migraines. They also discovered that the toxin stopped facial sweating, and began experimenting with injections of the sweat glands in the hands and armpits of excessive sweaters. The procedure works, but requires thirty to sixty injections every three to six months and costs an average of $1,200 per treatment.

Along with rank armpits, millions grapple with "feminine odor" and foul-smelling feet, and a huge industry churns out arsenals for battle. In hygiene-obsessed America, combating malodors has for many turned into a profitable career. Karl Laden, who holds a Ph.D. in chemistry, has advised the health and beauty industry on antiperspirants and deodorants for more than thirty years. He has written two volumes on their use and technological development; they serve as a sort of bible of B.O. for the personal products industry worldwide. Laden now works at a company he cofounded in Israel called InnoScents, which is working to develop new methods of fighting body odor.

Another such entrepreneur is Herbert Lapidus, an industrial

chemist who came up with the idea for Odor-Eaters, the latex inner sole with activated charcoal. Like other sweaty parts of the body, feet begin to reek when bacteria feed on perspiration. Because feet are trapped in socks and shoes all day, bacteria have the perfect conditions in which to grow, especially since each foot contains 250,000 sweat glands and can produce a pint of perspiration in a day.

As the head of research and development for a small company called Combe International, based in White Plains, New York, in the early 1970s, Lapidus was seized with the idea of solving the problem of foot odor. (He demurs at first from saying why, then says that his wife suffered terribly.) He saw a British invention, a paper sole that covered a layer of activated charcoal, which gave him an idea. You couldn't very well put paper inside a shoe and expect it to last, but the activated charcoal was a brilliant notion. Activated charcoal filters chemicals and odors in everything from nuclear waste plants to airport ceilings (the fumes would be intolerable without it). If you put the activated charcoal inside a bed of latex, he thought, it might work.

In order to keep the feet dry, Lapidus designed the latex soles with hundreds of tiny holes, which would act as a bellows with each step, circulating air. The charcoal absorbed the perspiration and the odors, and an antibacterial powder killed the bugs that led to bad smells in the first place. For six months, Lapidus worked around the clock to perfect his new sole.

To test the product, Lapidus put an ad in the *New York Times* advertising a remedy for foot odor. Overnight, he had dozens of willing participants. Each had stories more outrageous than the last. One woman refused to let her husband indoors until he had "aired out" his feet; his shoes had to be kept a certain distance from the family noses at all times. Another woman claimed to use a gas mask when she washed her husband's socks. One desperate wife said she was contemplating leaving her husband if she couldn't find a way to deal with his foot odors. He marketed the product as a one-size-fits-all, with patterns people could trim to fit their own shoes.

Lapidus is retired now, and declined to release sales figures. But he does say that if every pair of Odor-Eaters ever sold were placed toe-to-heel, they would circle the planet twice. According to market analysts, Combe's share of the odor-eating market is 85 percent, which would account for about $50 million a year in sales. Lapidus designed and patented many other products, including Lanacaine, Vagisil, and Just for Men haircolor. "But all anyone ever wants to discuss is Odor-Eaters!" he says. "I'm sure it'll be on my tombstone."

Then there is Mel Rosenberg, a microbiologist in Tel Aviv who is a leading expert on bad breath. Patients visit his clinic from around the world—some are so convinced their breath reeks, they are suicidal. One woman was desperate for help with her halitosis, but Rosenberg couldn't smell a thing. He quietly probed the cause of her insistence. As it turned out, the woman's father had always had foul breath, so she subconsciously assumed that she did, too. Eventually, she brought her eighty-four-year-old father from New York, and Rosenberg treated him. Often, he says, it is those with the least awareness of their odors who smell the worst. Another woman who had suffered from halitosis for decades was found to have a calcified bead lodged in a sinus. It had been there since she was three.

Like bad odors elsewhere on the body, bad breath generally comes from bacteria. If you don't brush and floss daily, particles of food remain in the mouth, collecting bacteria, which can cause bad breath. In addition, food that catches between the teeth, on the tongue, and around the gums can rot, leaving the odor of decay in the mouth. Dentures that aren't cleaned properly can also trap food and attract germs. Other culprits are foods in the *Allium* family, such as garlic and onions, which, once they are absorbed into the bloodstream from the digestive tract, are transferred to the lungs. The odor is expelled with exhalations. No amount of brushing or flossing can help until the body eliminates all traces of the offending food. Dieters, too, can get bad breath from not eating enough. When dieters burn stored fat, it gives off acetone, which has a rank, medicinal odor.

Though Americans spend a fortune on breath fresheners, they don't do anything but mask bad odors, Rosenberg says. "You have to kill the germs in order to get anywhere." Working with dentists, Rosenberg developed a two-phase oil and water mouthwash, which is on the market in Israel and the UK. Bacteria and debris attach themselves to the solution and are discarded.

The topic of body odors, from wherever they emanate, makes people laugh. It also makes them nervous. But obliterating our smells has become a national reflex. Consider the following, from a pamphlet distributed to foreign students at the Rochester Institute of Technology:

> . . . Americans have been taught that the natural smells of people's bodies and breath are unpleasant. Most Americans bathe or shower daily (or more often if they engage in vigorous exercise during the day), use an underarm deodorant to counteract the odor of perspiration, and brush their teeth with toothpaste at least once daily. . . . They rinse their mouths with a mouthwash or chew mints in order to be sure their breath is free of food odors. . . . Most Americans will quickly back away from a person who has "body odor" or "bad breath." This backing away may be the only signal that they are "offended" by another person's breath or body odor. The topic of these odors is so sensitive that most Americans will not tell another person that he or she has "bad breath" or "body odor." Some foreign students and scholars come from places where the human body's natural odors are considered quite acceptable, and where efforts to overcome those odors, at least on the part of men, are considered unnatural. Still other students and scholars come from places where personal cleanliness is considered more important than Americans and they may view most Americans as "dirty."[4]

Today more than ever, Americans are urged to embrace diversity, and to accept differences in race, beliefs, and religion. Muslim women may wear veils; Indian women may wear saris. But woe be to the newcomer who smells like one.

eleven

From the Bronx to America: Odyssey of a Nose

> I imagined that there was no happiness on earth for a man with such a wide nose, such thick lips, and such tiny gray eyes as mine. . . . Nothing has such a striking impact on a man's development as his appearance, and not so much his actual appearance as a conviction that it is either attractive or unattractive.
>
> —Leo Tolstoy

> A beautiful nose will never be found accompanying an ugly countenance.
>
> —Johann Kaspar Lavater

For as long as he can remember, Richard Garvey has been thinking about noses. Such awareness was born of many experiences. The first was an unfortunate collision between his nose and a stickball bat when he was in first grade (his nose was broken). The second came about a year later, when, one day after school, he sat transfixed before the television. As he watched Superman fly through the walls of a skyscraper to rescue Lois Lane, his mother interrupted to ask if he could go to the corner for some milk. "Sure, Ma," he remembers saying. "Just wait till the commercial." If he ran fast enough, he knew he could get back before the episode restarted.

Inspired by his hero, he bounded down the stairs and onto the street. A huge oak, encroaching onto the sidewalk, gave little Richie an idea. "I was learning in church that faith could move mountains," Garvey says, "and the damn thing was in my path." He thought that if he prayed hard enough, he could soar through the tree, just like Superman. So, clutching his quarters, he started to run. He picked up speed, stretched his arms before him, and flew—straight into the trunk of the oak. A woman waiting for the bus shouted, "Sonny! Didn't you see that tree?" Blood was everywhere—on his shirt, his face, in his mouth. It took a moment for him to realize that it was pouring out of his nose. Sobbing, he limped back home. Not only had he missed part of "Superman," his nose had swollen "like a zucchini," Garvey recalls. "My mother didn't know what the hell happened."

His nose healed, but it was badly broken. Never small, it now had a big hump, right in the middle. He could see it even without looking in the mirror. When you have a large nose, the bridge of the appendage is always within your field of vision, like the emblem of a Mercedes-Benz on the hood of a car.

In time, he grew both to hate, and to love the bump. It wasn't exactly pretty, but it was distinguished, differentiating. And when he felt insecure or nervous, he would rub it with his forefinger, almost as if for good luck.

Garvey had other reasons to care about noses, and not just his own. The youngest of six children, Garvey helped to care for his sister Valerie, who had spina bifida, lung and heart defects, and severe facial deformities. The two were the closest in age, barely two years apart. When he was very young, his mother took him aside and asked him to help her. "She's a special child," she told him. "She's not going to be with us very long, and God wants us to take extra special care of her."

Garvey's father died when he was two, and an older brother was grown. So he assumed the role of Valerie's caretaker as a boy of five. He drew her blood for tests, and gave her medicine from a dropper at bedtime each night. "Kids at school would make fun of her, call her names, and I would just get furious," he says. The girl's nose was small and misshapen, and her voice was high-pitched. "They'd laugh, because she looked different, sounded different. I would storm out of the house and want to beat up any kid who made fun of her. I even hit somebody once in church." She died at thirteen.

A few years later, Garvey was again watching TV, and as he flipped through channels, he saw a PBS special on plastic surgery called "A Normal Face." He was awestruck. On the show, plastic surgeons repaired cleft palates, aligned asymmetrical eye sockets, and grafted new skin on the cheeks and noses of burn victims. "I always thought plastic surgery was for women of a certain age, or people who were in the federal witness protection program," he says. (His neighborhood was thick with wiseguys.) Now he wondered what Valerie's life would have been like if she had had such an operation. "Maybe we could've helped her have a more normal life," he says quietly. "Maybe people wouldn't have been so cruel." He decided then what he would do with his life, despite other distractions.

And so at fifteen, he started saving for his education by working at a local restaurant near Yankee Stadium. Signed pictures of Joe DiMaggio and Mickey Mantle line the walls. From busboy to bartender, Garvey worked weekends, nights, and summers to put himself through school—first through Columbia University, and then through medical school at Georgetown. Through it all, he says, his work at the restaurant helped remind him who he was, and where he came from.

But for his mother Lorraine, at least some of where he came from was worth forgetting. As the daughter of Sicilian immigrants in the Bronx in the 1930s and 1940s, when anti-immigrant sentiment was high, Lorraine had claimed that her surname, La Forge, was French. Once she married Al Di Piero, though, there was no doubt about her or her family's identity. Still, when Richard was young, she took him to speech therapy in order to take the edge off his Bronx accent. "You have to be totally American to succeed here," she told her son. Richie Di Piero understood that the chance of his somehow metamorphosing into Greg Brady was slim. But some years after his father's death, his mother married Martin Garvey, an Irish-American bus driver. He adopted all the Di Piero children and bestowed, along with love and stability, what was to become an extraordinary gift: an ethnically ambiguous surname.

"You're a real American boy now, Richie," Garvey's mother told him, "with a nice American name."

But still, there were always reminders that somehow he was

not. For one, the nose. Second, his accent, though diminished, remained. And Columbia's wealthy elites did their best to remind him of his origins. "There were guys with Roman numerals after their names, and they'd give me all kinds of shit. 'Hey, here's our smart little wop! Hey, Richie, say sometin'!' " Garvey winces. There is nothing worse, he says, than someone from Connecticut trying to mimic an outer borough accent.

It was hardly as if he were trying to be someone else in the first place. He points to the walls in his basement, covered with its own icons: the all-Italian cast of *A Bronx Tale,* and the nearly purebred crew of *The Godfather.* A signed picture of Robert De Niro hangs above the fireplace. But still, the comments stung. He taps his nose. "They stuck with me, you know."

Training for plastic surgery is long, beginning with four years of medical school and a residency of seven years. But Garvey was determined to enter the specialty even as he enrolled at Georgetown. He knew firsthand how a nose—"a nose!" he says, incredulous at the thought—could mark a person, and before long he was a man obsessed. He became fascinated with their shapes, their anatomy—and how to change them. In his spare time, he watched doctor after doctor sculpt and streamline noses, and he made notes to himself about whose surgical style he would eventually emulate.

What surprised him most, though, was learning about the long history of rhinoplasty, as nose jobs are formally known. As it turned out, people's concerns about "fitting in" on account of their unusual noses went back centuries. Plastic surgery takes its name from the Greek *plastos,* meaning molded or shaped, and while there were reports of repairing wounded features in Asia, the most detailed versions did not appear until the Renaissance in Italy. Most celebrated was the work of Gaspare Tagliacozzi, a surgeon and anatomist from Bologna who described a method of substituting noses lost to injury or disease with the skin of the arm. He published his findings in a book, *De curtorum chirurgia per insitionem* (The Surgery of Defects by Implantation), printed in 1597. The Church was aghast at Tagliacozzi's work—officials believed it to be tampering with the work of God—and excommunicated him. At any rate, his technique was short-lived, as the reconstructed noses tended to freeze in cold weather, or get dislodged during a violent sneeze. But Tagliacozzi knew the benefits of reconstructive surgery were also psychological. While the new features lacked physical perfection, they helped to "buoy up the spirits and help the mind of the afflicted."[1] Elsewhere in Europe, people experimented with other nose replacements. After losing part of his nose in a duel, the Danish astronomer Tycho Brahe fashioned a mixture of metal and wax, which he affixed to his scar.

In India two hundred years later, a British officer noticed the peculiar nose and a large scar on the forehead of an Indian merchant with whom he traded. The officer learned that the man's nose had been sliced off as punishment for adultery. A potter who was trained in crafting new noses had made him a new one of wax, which was placed on the stump. A flap of skin was cut from the forehead and grafted over the prosthesis. The technique, which had been used for centuries on the subcontinent (cutting off noses was a common penalty), was soon reported in the *Gentleman's Magazine of London* in 1794.[2]

The article attracted great attention. Europe was in the midst of a syphilis epidemic, and noses were obvious casualties of the disease. The infection attacked the soft tissue of the nose, causing the bridge to collapse, or worse, leaving a gaping hole in the middle of the face. It was a public marker of private behavior, even branding sons and daughters with the sins of their parents (infected mothers could pass the disease to their babies). A nineteenth-century German surgeon devised surgical tools and materials ranging from wax to gold to rebuild sunken syphilitic noses, and an American plastic surgeon once butchered a live duck in the operating room so he could graft the bird's breastbone to the nasal cavity of a syphilitic man.[3] However, incisions to insert the new materials, which later included ivory and glass, were made on the outside of the nose. They were highly visible, but they could at least help those with disfigured features slip more easily into society.

More than anything, early plastic surgery involved reconstruction of the nose. But by the late 1800s, there was a shift from restorative to cosmetic surgery as immigration—and xenophobia—transformed the modern world. From Jews in prewar Germany to Irish immigrants in England and the United States, noses had a way, much as they did in Lavater's day, of branding—however wrongly—a person's character. John Orlando Roe, a surgeon in Rochester, New York, performed an operation to "cure" the Irish "pug" nose. Borrowing from Jabet's classifications, Roe concluded that the pug nose was the outward sign of weakness and stupidity. Roe crafted straighter, more linear noses, and was the first in the United States to use internal incisions, making cuts on the inside of the nostrils so as not to leave evidence of his work.

Shortly thereafter, an orthopedic surgeon in Germany, Jacques Joseph, made a similar breakthrough. Born Jakob Lewin Joseph to a rabbinical family in Königsberg, Prussia, Joseph changed his biblical name to the more cosmopolitan Jacques as a student in Leipzig and Berlin. During World War I, new explosive devices inflicted facial casualties on a mass scale, but improved hygiene and sanitation helped keep alive wounded soldiers who would otherwise have died of infection. Advances in anesthesia had also helped to make surgery more commonplace.

Joseph reconstructed the faces of countless German servicemen at a Berlin hospital. A dour man who bore dueling scars on his cheeks—a mark of prestige and acceptance into German society, especially for a Jew—Joseph devoted himself to studying the repair of wounds and deformities. He devised tools for many of his procedures, some of which are still in use today.

Like Roe, Joseph also concerned himself with transforming faces that advertised ethnicity—particularly those of his fellow Jews. In 1898, an elegant, wealthy merchant approached Joseph for a consultation. On Sunday afternoons, he customarily drove his family through Berlin in their stately carriage. But something held him back from fully enjoying his outings: his nose. Though he tried to offset it with a thick black mustache waxed into upward curls, the man still believed that his prominent profile drew unwanted attention to his family. They might invite fewer stares if only his nose was smaller. Could it possibly be changed? the patient asked. Joseph assured the man that it could. In surgery a few weeks later he filed down the cartilage on top of the man's nose, which resulted in a new look with "gentile contours."[4] The gentleman returned to the driver's seat of his carriage, his nose transformed. Word of the change spread swiftly through Berlin, and Joseph soon had a burgeoning cosmetic practice.

Indeed, between the world wars, anti-Semitism was a growing feature of life for German Jews. In an increasingly hostile culture, many Jews—some whose families had been in Germany for centuries—began to fear for their jobs and their futures. If altering a small matter—a distinct nose—could make matters easier, such a change was welcome. And since Joseph also made incisions on the inside of the nostrils, he left no trace of his handiwork. (In fact, Joseph thought he was the first to use this method until he learned of papers by Roe.) Many Jews underwent the procedure at the hands of a surgeon they found sympathetic to their concerns, and Joseph later said that he was only too happy to help those he found to "suffer a Jewish nose." He performed many such rhinoplasties free of charge.

Joseph promoted the psychological benefits of cosmetic surgery, writing in his 1931 book *Rhinoplasty and Other Facial Plastic Surgery,* that a person's well-being depends, to a large degree, on his sense of confidence and dignity. Such attributes are enhanced, he wrote, when facial features are in harmony. Joseph's book included a nasal "scale," which touted as its ideal a woman whose nasal tip was no greater than 30 degrees from its bridge. (To some extent, his theory persists today.) Joseph believed that anyone self-conscious about his or her features deserved to feel more self-assured—particularly as Nazism gained currency. If a smaller nose could help a person feel better, Joseph reasoned, it was bound to have some functional benefit.

After World War I, Joseph's reputation as a reparative surgeon was widespread. But, as he performed more and more rhinoplas-

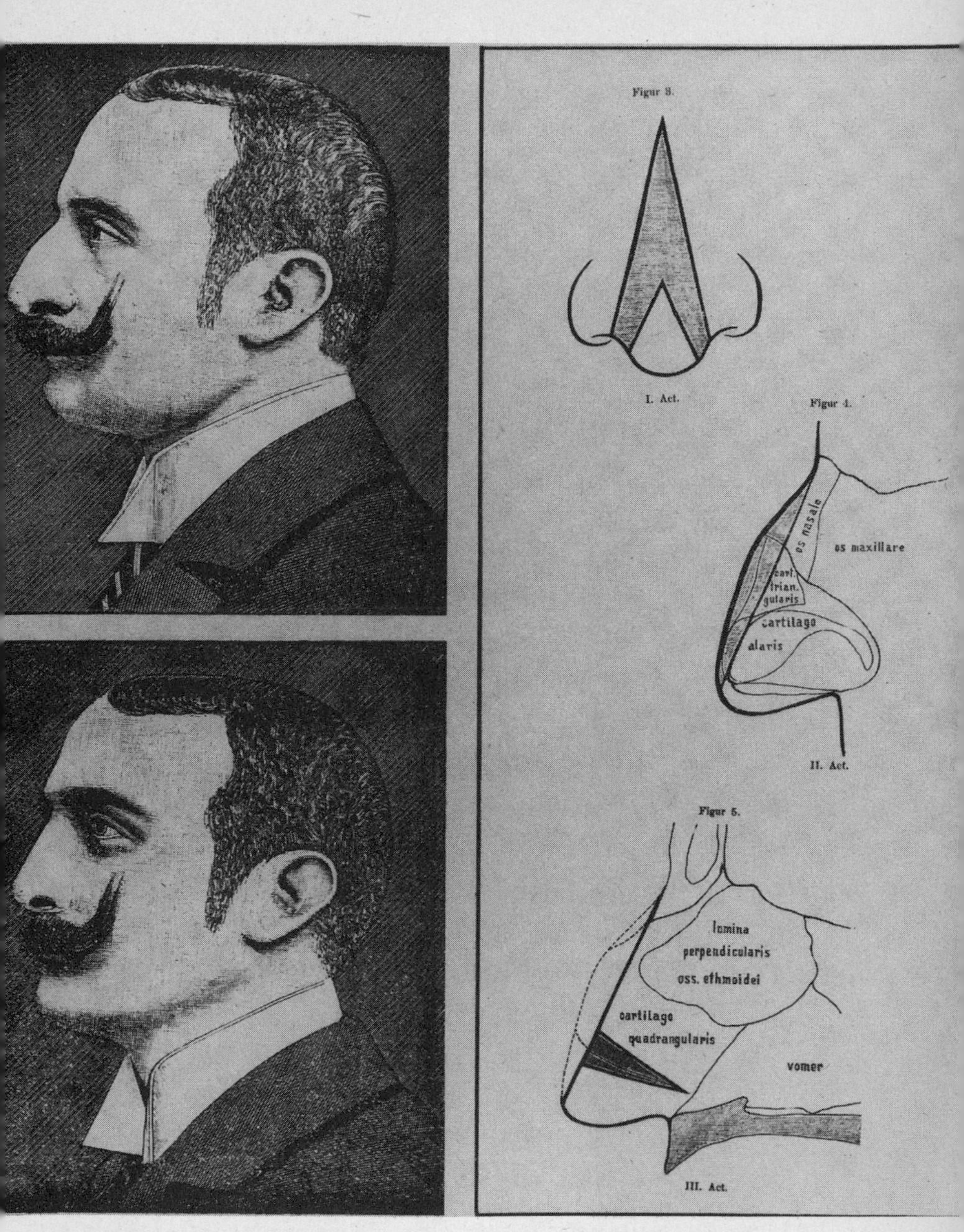

Joseph's pre- and post-operative diagram of an early rhinoplasty patient.

ties, that regard soon slipped to scorn. Medical officials, concerned that he was performing procedures for vanity's sake (or worse, in order to camouflage Jews), revoked his operating privileges throughout the city. Whether in principle or out of fear of increased scrutiny, even Berlin's Jewish hospital barred him from performing surgery. Finally, he opened his own clinic, which drew curious surgeons and apprentices from abroad.[5] A German patriot, Joseph died in 1934, just months after Hitler seized power.

The year after Joseph's death, the Nazis passed the Nuremberg laws, which reserved citizenship in the Third Reich for Aryans alone. Jews were stripped of their German citizenship, forbidden to marry Germans and hire German domestics. More than ever, Jews who remained in Germany felt the need to become as inconspicuous as possible. Nazi schoolteachers began to preach the "principles" of racial science put forth by Adolf Hitler. They measured skull size and nose length, and recorded the color of their pupils' hair and eyes to determine whether students belonged to the true "Aryan race." Jewish and Romani (Gypsy) students were often humiliated in the process. Soon enough, those lacking "Aryan" features sought to change prominent noses by surgeons familiar with Joseph's techniques.

Postwar America: Joseph's Legacy

In postwar America, where many of Joseph's protégés had taken up practice, plastic surgery flourished. The timing was perfect.

Americans trying to conform to strict standards of beauty sought "nose bobs," as they had come to be known, in huge numbers. Joseph's legacy was to change the shape of the American nose for generations.

Joseph's disciples guarded their techniques carefully, and as rhinoplasties became more popular, this presented a problem: not enough doctors knew how to perform them. Samuel Foman, an anatomist and medical textbook writer, had traveled to Germany before Joseph's death and had a good grasp of his methodology. But in order to describe the procedure in textbooks in as much detail as possible, he needed to observe it. Yet whenever he entered an operating room, American plastic surgeons covered their work. Undeterred, Foman sought to teach the techniques to otolaryngology students, who, he believed, understood the nose as well as plastic surgeons, if not better.

Among his students at Mt. Sinai Hospital was a brash young ENT named Irving Goldman. He shared Foman's interest in the nose's anatomy—so much so that on his honeymoon, he traveled to Germany to study with a Hungarian doctor trained by Joseph, Zoltan Nagel. Eventually Goldman became well known in Manhattan, and refined a rhinoplasty technique known as the "Goldman Tip," which narrowed the bridge of the nose, or *dorsum,* and divided the cartilage that formed a dome over the tip. The result was a pert little nose. Goldman believed his procedure gave ideal support to the new noses he shaped, which, after removing large amounts of cartilage and bone, were radically different features.

The stylish new nose was helped along by another innovation: television. Women were already eager to emulate their perky idols; those who lacked genetic luck could always get a bob to look like Judy Garland, Shirley Jones, and Debbie Reynolds. Between Annette Funicello's years as a Mouseketeer and a bikini-clad sidekick to Frankie Avalon, she altered her wide nose to match her frisky new all-American image. Not only wholesome actresses shared such features. There is great debate about whether Marilyn Monroe had rhinoplasty or not, but at any rate, she bore a small, upturned nose, much like the one Barbie dolls would sport some years later. (Not everyone fit the pattern—or wanted to. Despite the advice of many, Sophia Loren refused plastic surgery, preferring instead to keep her long nose. For a time, American photographers were forbidden to snap her in profile, but the ban was relaxed as her appeal spread.)

But for those of long nose who lacked Loren's curves, the pressure to look a certain way was great. Harold Holden, a Los Angeles plastic surgeon who trained with Joseph, wrote of the gratification he derived from "fixing unaesthetic noses." He describes one patient, a teenager, who was so disturbed by her large nose that she wanted to quit school. Though a talented singer, she nevertheless avoided performing—she was convinced people would ridicule her profile. She despaired that her looks disappointed her parents, who, from time to time, lamented that she lacked a "cute nose" like her older sister, June. "I can't go on. . . . since Dad has become pretty well-known in his field and he and Mother have stepped up our social life, I feel more than ever that I am the scarecrow of the family, and I've gotten to where I just don't want to be seen."

Holden agreed with the girl's assessment of her nose, writing that it was "pretty bad." Such misfortune would surely be the girl's ruin. "This is a situation almost anyone and especially a young girl would find it hard to stand up against. This is a youngster whose whole future is at stake." Holden thoroughly reshaped her nose. "From the moment the final dressing was removed, the girl's happiness was a thing to see. Her deep feeling of inferiority to her sister was both revealed and ended in her first words: 'Oh, Doctor! Now my nose is not in my way! It is even better than June's!' " Holden, too, was pleased. "A life had been weighed down with deformity. 'Burial at sea' under waves of misery appeared to this adolescent personality the sole possibility in life. When the deformity was removed, the girl became able to cope with all the normal currents of her existence."[6]

Another doctor who became well-known for his transformational powers was Howard Diamond, a Manhattan plastic surgeon who perfected a scooped-out bridge and upturned tip that was popular up through the 1960s and early 1970s. According to some of his former students, Diamond was a veritable nose-bobbing machine, performing as many as five operations per morning.

Like Garvey's mother, many second-generation American parents were eager to give their children every possible advantage. Many followed their moves to the suburbs with offers of nose jobs for their daughters, just in time for their bat mitzvahs or sweet sixteen parties. But by the mid-1960s, style—and times—were changing. Revolt overtook conformity from politics to pop culture. Bob Dylan's antiestablishment lyrics, reedy voice, and less-than-classical looks helped underscore his appeal. Though Ringo Starr was dubbed "The Nose," attention to the Beatles focused on their music and spiritual quests, not teeny-bopper beach frolics. Barbra Steisand leapt from her role in *Funny Girl* to leading lady opposite Robert Redford in 1973. Her popularity soared. Beverly Johnson, an African-American model, appeared on the cover of *Vogue,* displacing decades of Grace Kelly look-alikes. Cher, too, became a national icon (although, unlike Streisand, she eventually got her nose fixed. Streisand rejected the idea of altering her nose for fear of changing her voice).

What was essential for parents reared in the 1940s and 1950s—a strict set of rules applying even to appearance—rarely made sense to their children. Outlandish and "natural" replaced the docile and the familiar. From long, free-flowing hair to the Pill and legal abortion, young women suddenly had choices about their bodies that were unimaginable to their mothers. The primer of feminist acceptance, *Our Bodies, Ourselves,* went from an underground publication in 1970 to a paperback tome that ultimately sold 4 million copies—and stood out openly on shelves from college dorm rooms to mainstream bookstores.

The number of people getting nose jobs, meanwhile, dropped precipitously.

By the late 1970s, not only young women were questioning the wisdom of radical rhinoplasties—doctors joined in as well. The procedure was still the most common cosmetic operation, despite the era, and few plastic surgeons were sitting in empty offices. In fact, women who had their noses hollowed out and upturned began showing up with myriad complaints. It can take as long as two years for a rhinoplasty to heal fully, and many noses that had been completely transformed began to look odd once tissue scars settled and skin shrank around the new nose. Some of them were misshapen, even frozen, on faces with otherwise strong ethnic features. Suddenly, doctors began to ask questions:

Why did this occur in this patient but not in that one? What hadn't been done that should have been—or, more pointedly, what had been done that shouldn't have been?

And as the culture of plastic surgery evolved, so did technology. In the 1960s, plastic surgeons had trained by watching their mentors in the operating theater, often with binoculars. But by the early 1980s, young doctors didn't even need to leave their offices—they could watch the greatest practitioners on video. More than anything, this advance changed the profession, says Robert L. Simons, a Miami facial plastic surgeon who acted as president of the American Academy of Facial Plastic and Reconstructive Surgery in the mid-1980s. Students—and teachers—became much more willing to question procedures and outcomes as they stood in front of a VCR than if they were inside an operating room. And suddenly students began learning from people all over the United States, not only the doctors they had observed during their residencies. One doctor's technique for fixing a certain problem might well be incorporated with that of another. Such openness encouraged physicians to be more considerate of their patients, and as a result operations became more individualized. But since the days of Joseph, the general procedure has been roughly the same. The surgeon cuts inside the nostrils and separates the skin of the nose from the underlying bone and cartilage. The bone and cartilage are then reshaped and the skin is redraped over the surface. Occasionally a surgeon will make an incision in the skin between the nostrils, allowing him to more easily see—and reshape—the nasal tip.

By the mid-1980s, most plastic surgeons had dispensed with

the notion that they "knew best," and began asking patients what they, in fact, wanted to change. Doctors stopped turning the noses on angular faces into "cute" features. Even the terminology changed. Cartilage was left in place, and tips were "refined" or "reshaped"—not named for a surgeon who honed a certain technique. "By doing a little," Simons says, "we learned we could achieve a lot."[7]

But elsewhere, a deft touch was not necessarily the rule. By the late twentieth century, cultures from Brazil to Japan pursued perfection from the knife—at any cost. In Rio and São Paulo, where large derrières are a hallmark of beauty, even the poorest of women rounded their rumps with buttock implants. In Tokyo and Seoul, teenagers and young women sacrificed to pay for operations that would dramatically "Westernize" their faces. Two procedures are highly popular: one removes skin from their eyelids, while another raises the bridge of their nose (for those who aren't entirely convinced, removable nasal implants, which slide up into the nostrils, are available).

Social scientists say that beauty everywhere relies on symmetry. Even in the most remote societies, the loveliest woman and the most handsome man have balanced features and shapely bodies. But notions of beauty also depend on the prevailing social mores. Nowhere is this more evident than New York City, where the idea of the perfect nose can vary from neighborhood to neighborhood. In Manhattan, which shares a perch with Los Angeles as the capital of American plastic surgery, there are even a handful of experts for each ethnic group.

Until recently, though, plastic surgery textbooks prescribed the classical noses Da Vinci depicted five hundred years earlier—the "ideal" faces of Northern European noblemen. While many patients still seek the delicate silhouette of Anglo-Saxon features, they simply don't belong on many Mediterranean or non-Caucasian faces—which means the majority of the population in many of large American cities.

Some years ago, Ferdinand Ofodile, the head of plastic surgery at Harlem Hospital, began to worry about some of the requests he was hearing. While many in the African-American community recoiled at the sight of Michael Jackson's whitening skin and ever-dwindling nose, some patients approached Ofodile with pictures of white entertainers whose features they wanted to emulate. "The great majority confessed to hating their noses," he says. "It was as if the negroid nose was something that had to be changed in order to be acceptable. For many, to be born with such a feature is by definition to have something wrong."

An authority on the treatment of black skin, Ofodile traces his roots in medicine in the Ibo country of Nigeria for generations. (African skin can turn black, white, or form fibrous keloid tissue after it is injured or cut in surgery.) But Ofodile was also drawn to the nose, and the rich variations among black features worldwide. Unlike the Jewish and Italian teenagers who got nose jobs a generation earlier, many of Ofodile's patients were Latino, Caribbean, and African-American adults. Ofodile, an elegant man who wears French cuffs and speaks with precise diction, realized that few in the rarefied world of plastic surgery understood the differences in black features—and how to change them subtly.

"It can be distressing for someone to feel trapped inside their body, or their face," he says. "To affect such change in a positive way can be very gratifying. But a face—this is something that must be approached with the utmost sensitivity." And so Ofodile began publishing articles in plastic surgery journals to educate fellow surgeons about black noses. "The history of plastic surgery is seen largely through white eyes," he says.

First, he sought to explain nose structure as a matter of evolution. Because the nose's principal functions are to warm, cool, and moisten the air drawn into the lungs, its shape evolved to reflect its climate. And so, before the dawn of mass migration, the length of noses of those in cold, dry climates became long and had a narrow base, which increased the surface area and the period of time over which inspired air was warmed and humidified. Conversely, in hotter, muggier climates, the nostrils became more circular and had a wider base, decreasing the surface area and the time over which warming and moistening of a breath took place.[8]

Through the volume of patients he saw, Ofodile created three categories of African noses, which other surgeons soon used as a model: African, Afro-Caucasian, and Afro-Indian. Many with African noses sought Ofodile's help for flat bridges and wide nostrils; those with Afro-Caucasian and Afro-Indian noses wanted to change drooping tips or excise humps from the dorsum, the part of the nose that extends from the bridge to the tip. But mostly, Ofodile sculpted new, more defined tips, and performed augmentations: increasing the size of the dorsum by grafting cartilage from elsewhere, usually an ear.

Slowly, as surgeons began using more restraint among whites, those with ethnic patients began to scale back the procedure as well, tailoring it to other features. Since the late 1980s, Ofodile says, rhinoplasty has been the most popular facial plastic procedure among American blacks. African-Americans now make up 7 percent of patients seeking plastic surgery, up from 4 percent in 1992. Latinos account for 10 percent, and Asians 4 percent. Individual procedures are not broken down, so it is impossible to tell the numbers for rhinoplasty itself.

At about the time those categories were published, Richard Garvey was starting his residency in his old neighborhood, at Jacobi Hospital, a public facility in the Bronx across the street from his grandmother's old house. Most of his first cases were reconstructions: burns, trauma, gunshot and knife wounds. Soon he came to the same conclusion Ofodile and Simons had, that plastic surgeons needed to respect the patient's wishes and individuality. "Plastic surgeons weren't God. They had skill. But it was nothing without listening, and looking at your patient."

And Garvey had many patients to listen to, in many different languages. Like other major cities, New York—and particularly the Bronx—was becoming a more diverse place. Gone were many of Garvey's "paisans." Italians and Jews, once the bulwark of the borough, had long since moved to the suburbs, replaced by Latino and Caribbean immigrants who presented challenges unlike those he had encountered in the lush, leafy world of Georgetown, where senators came for "stealth" facelifts.

At Jacobi, most of Garvey's first patients were Latino. Many were gang members, and bore not only slashing scars on their cheeks but whole chunks of tissue missing from the dorsums of their noses, also a result of run-ins with switchblades.

One wintry day Garvey prepared to reconstruct the nose of a seventeen-year-old boy, Ricardo, who had recently dropped out of high school. Ricardo sat nervously in the patient's waiting room, twisting his hands and then wiping his sweaty palms onto his jeans. Deep gashes lined both sides of his face. A scraggly beard covered sore-looking pimples. Garvey had described the wounds on the boy's nose, but from the front, they didn't look that bad. When Ricardo turned his head, however, he looked like the famous painting of the Duke of Urbino that rests in a room of its own in the Uffizi. A finger-sized scoop near the bridge of Ricardo's nose was missing. The sight of it was alarming, to say the least. Garvey mentioned "harvesting" cartilage from deep inside the septum. I had skipped breakfast, and suddenly I was glad I had.

"You doin' OK, man?" Garvey asked. "We're going to fix your nose, pal. You'll breathe better, and you won't be upset when you look in the mirror, OK? You gotta just promise me to stay out of trouble, OK?"

The boy, seventeen, looked on, dazed. "I'll try, Doc, I'll try," he said.

"You clean, man?" Garvey asked. Ricardo nodded and crossed his heart. "I'm clean, Doc, clean." (Cocaine users are at special risk during surgery because the drug can seriously damage the lining of the nose and can cause intense constriction of the blood vessels. "You can hemorrhage," Garvey told me matter-of-factly. "It's bad. In the eighties, sometimes addicts died in the OR.")

In the operating room, Ricardo was stretched out like a dead man, arms motionless at his sides, toes poking through the end of the blankets. Garvey started to inject his nose with a numbing agent; even though Ricardo was asleep, he didn't want to risk his feeling anything. Garvey's partner, Don Roland, reached into Ricardo's nostrils with fine long scissors to snip his nose hairs. Garvey then made incisions alongside the septum with a tiny scalpel, and urged me to peer inside. Blood spurted from each cut, and the doctors moaned knowingly: too much coke in the days before the operation.

Garvey peeled the skin back on the septum, and began to operate. It was stiflingly hot under the lights, and it felt very crowded. Garvey took a minute instrument that resembled a cheese slicer, and muttered what sounded like a prayer. The plan was to take cartilage from deep inside the septum, then graft it onto the hole in Ricardo's nose. Beads of sweat appeared on Garvey's forehead, and even though I'm not squeamish, I felt faint.

With a grunt and a stunning show of strength—Garvey is a big man, with powerful arms—he forced the blade of the device deep into Ricardo's septum. The violence of it was striking, and the room fairly snapped with intensity. Everyone waited nervously to see what he would retrieve. Finally, he held up a small piece of white tissue. It looked exactly like the cartilage on a chicken drumstick.

Garvey tucked it into the hole, and started sewing Ricardo up. Roland clipped the sutures as short as possible. "Otherwise," he said, "they'll itch." Even with tubes in his mouth and his nose still swollen from the force of the operation, Ricardo looked wildly different. His deformed nose had skewed his whole face. Now, he looked handsome—or at least as if he had a chance at it.

For several years, Garvey has practiced in the Bronx, at a suburban New York practice and in an office on Fifth Avenue in

Manhattan as well. His following is large, a fact he attributes to his honesty about his own flaws: a thickened waist, his humped nose. When patients come in complaining about every wrinkle and blemish, he tells them: "Look at me! I'm no symmetrical masterpiece!" His demeanor—and family history—play a role too. "I remember what it was like to be the one asking a doctor for help," he says. "I know all about the anxiety you feel when you're on the other side. You're thinking, 'Will this person be able to help me or not?' "

After seeing hundreds of satisfied patients, he began to consider changing his own trademark: his nose. What's the big deal? he thought. By 2000, rhinoplasty seemed downright routine. It had outpaced all cosmetic procedures in the United States, from liposuction to breast augmentation. Some 389,155 Americans had their noses reshaped that year, at a cost of $1.11 billion. Men accounted for 39 percent of the patients.[9]

The older Garvey got, the bigger his nose seemed (fatty tissue at the tip begins to sag with age). Garvey began to scrutinize his appearance more, and more, and more. He'd look at himself in photographs, smiling with his wife and three children or on the beach with old college friends, and all he saw was his nose. "I swear to God, it looked like a big ugly bird had landed on my face," he says. In his spare time, he once again started studying plastic surgeons—whose techniques he admired, who he thought "listened" well to his patients' requests, and with whom, he says, he felt "simpatico."

Finally, he decided on a surgeon, Mark Erlich. They were acquaintances, not friends. "God forbid I wasn't happy with it—then what would I do?" The day of the surgery, Garvey was nervous, and asked his wife and a friend to help him get home. Once in the holding area before the operation, he fretted. Was he really doing the right thing? Time and time he had counseled his patients: It's not so painful, and you'll be back on your feet in no time. Now, he himself was worried. What if the procedure was more than just "uncomfortable?" He had lived with his nose his whole life. Would changing it be worth it?

Hours later, he awoke in the recovery room, hysterical. He ripped the IVs from his arms, and insisted that he get up and walk. Two nurses and a very large orderly had to wrest him back onto the cot. "Let me out of here!" he screamed. "I want to go to my own bed!" He felt as if he couldn't breathe beneath his splint, and began gulping for air. He couldn't urinate, and he had an insatiable thirst. The sensations gave him a full-fledged panic attack. His heart raced, and he lashed out at everyone—nurses, his wife, his friend.

Once home, he was scarcely better. Clearly, he was having a bad reaction to his anesthesia: it had relaxed his bladder too much, created his thirst, and made him "nuts." He picked up the phone and, groggy, called Erlich, who was still in surgery and was unreachable. Then he dialed every anesthesiologist and urologist he knew in New York City for advice. One suggested taking a warm bath. Another prescribed Xanax, a tranquilizer, to calm him down. Finally, Erlich called. Garvey exploded, he says, sounding off like John Gotti. "If I don't pee in the next hour," Garvey screamed, "your anesthesiologist is going to be in the fucking East River!"

Armed with Xanax and Vicodin, a painkiller, Garvey finally settled down enough to sleep. Through it all, he dreamed of his patients, and the times they, too, had experienced such discomfort. "I felt so trapped and numb. Finally I realized what all my patients had gone through. It was just awful. I had never experienced anything like that—or been so out of control."

The following few days, Garvey was his own worst nightmare. While he gives his rhinoplasty patients strict instructions not to touch the inside of their noses—nostrils are lined with stitches—he couldn't help himself. He picked at the blood clots until the sutures became loose, and one by one, dislodged entirely. Then he began to blow his nose—hard. The more he did it, the better he felt, even though he knew he risked blowing air into the newly stretched skin around the outside of his nose. One day he blew so violently that he felt air travel beneath his eye and cheek all the way up his forehead. He dispatched his wife to the drugstore for Afrin—a topical (and caustic) decongestant that stung the tender, swollen tissues.

Meanwhile, Erlich couldn't believe what he was hearing. "Is this guy insane?" he wondered, and called Garvey's mentor at Jacobi, Bruce Greenstein. "I hope he doesn't manage his patients like this," Erlich told him. Greenstein assured him that as a surgeon, Garvey was as conservative and caring as he had ever seen.

But as a patient himself, he was a "self-destructive machine." One night, unable to breathe, or sleep, he bolted out of bed and pried off his splint—three days before it was supposed to come off. Relieved, he settled down to watch *Goodfellas,* his favorite movie. He still felt constricted, though, and went to the mirror once more. There, he peeled off the tape—again, days away from its scheduled removal.

A week after the surgery, he walked sheepishly into Erlich's office—no splint, no tape, no stitches. Erlich just shook his head. "I hate working on doctors, and operating on plastic surgeons is nothing I ever look forward to," he said. "But you are the worst patient I have ever seen." Indeed, Garvey had subcutaneous air bubbles beneath his nose, which Erlich had to aspirate with a syringe. He ordered Garvey not to blow his nose again for weeks.

But Garvey couldn't help himself. Once in his car, he took a handkerchief and blew, even before realizing he had. Too embarrassed to return to Erlich, he drove instead to his own office, where his partner removed the bubble once more. "Man, you're like a whale with a blowhole," Roland said. This time, the needle's insertion left a tiny scar on the bridge of his nose. "Serves me right," Garvey says.

Two months later, Garvey has had plenty of time to think about his new profile. He is thrilled to have the old bird gone, he says. But at the same time, he misses it. "Just a little bit," he adds. He doesn't feel nostalgic for the broad tip, or the cleft on the end someone once told him "looked like an ass crease." Yet he does miss his bump. "It had a strength to it," he says, with a twinge of sadness. And like today's newly shaped noses, it is, to be honest, scarcely noticeable. Garvey, broad and gruff, with sturdy features and dark, expressive eyes, has a nose that still matches his face—not that of a patrician, or one with an indented bridge. And the experience certainly deepened his understanding—and empathy—for his patients. It wouldn't be such a bad idea if every surgeon had to endure the operation he performs, Garvey says. (In fact, Garvey now describes all post-operative possibilities with his patients, including bad reactions to anesthesia. On the off chance they might react similarly,

Garvey prescribes Xanax. "It's unlikely that anyone will end up like me," he says, "but even so, I want my bases covered. I don't want anyone going through that if it can be avoided.")

Garvey says his new nose makes him feel more at home in the world. He no longer feels *beau-laid,* or "ugly-handsome," as the French would say. "An unattractive nose on an attractive face can be a huge distraction," he says, shrugging. "An attractive nose can take you to another level."

If someone meets him today, he says, they are no longer likely to think, oh, Rick Garvey, Italian guy with an Irish name, big nose. Not long ago, he attended a lavish wedding in the Caribbean, among a wealthy international set who, if they weren't born with perfect features, had certainly arranged to get them. For the first time, Garvey realized, he felt comfortable in a crowded setting of "mixed" people—in other words, not the native New Yorkers he counts as his intimates—Italians and Jews, all. There were wealthy American WASPs and European aristocrats. He could even take a sip of champagne without having to tilt the whole glass back beneath his nose. For the first time, it fit into a flute. "I looked around and realized, hey, man, no one is looking at your nose," he says. "It felt good. I felt, well, American."

But did a new nose really steady his soul? Were Tagliacozzi and Joseph right? Does a new appearance really help a person "fit in" better, or does one's perception of oneself merely change so that it feels that way? "I don't know," Garvey says. "But I can say that I don't miss the end of that big ugly snout one bit."

And yet Garvey realizes that there are certain taboos against discussing such feelings openly. While those opposed to plastic surgery dismiss the very notion of it as vain or "selling out,"

Garvey disagrees. People want beauty everywhere—in art, in their gardens, even their food. "When someone has a gorgeous home, you don't walk out saying, 'Man, what a vain person!' You say, 'Nice house.' As a plastic surgeon, you work on the surface. You fix scars or cut away the bumps or time, is all—it's like changing drapes that don't work. Of course I wish that the outer layer wasn't so important. But look at the Romans! The Greeks. The pursuit of beauty has been around a long, long time."

Mostly, he finds deep gratification in his work, in patients who tell him that he changed their lives. One, a woman in her thirties who was self-conscious of her nose, says she keeps him in her prayers nightly. She believed her nose kept her from having the confidence she needed to meet the right man. And there are trauma victims, such as the little girl whose nose and face were slashed by hundreds of pieces of glass when a tire fell off a truck and into her windshield. "You do everything you can," he says. "You just want people's lives to be easy, you know? Or at least not as hard as they have to be."

There are, of course, the demands of those who expect the impossible—women who have had numerous rhinoplasties and go from surgeon to surgeon in the hope of finding the "perfect" nose. Structurally, the cartilage and delicate nasal bones can only stand so much, and once in a while, a person will come in on their sixth or seventh procedure. "Those are the people you send away. You say, 'You can find someone in this town to help you, but it's not going to be me.' " Garvey looks up, hears his young son crying, and heads up the stairs. "Some people," he says, "think it will change everything about their lives."

That, he says, is impossible.

It will be a long time, of course, before the contours of exotic noses, like Garvey's "birth nose," as a woman I once met called her wide one, are celebrated by mainstream America. Unusual noses are suspect; strange tips and prominent bridges remain as unwelcome as garlic breath and "nervous B.O." An essay by an Orthodox woman in a Jewish weekly in New York went so far as to justify plastic surgery under Jewish law.

It would be nice to believe that in a time of multiethnic supermodels, worries like that of Holden's patient, or Richard Garvey's, or mine, perhaps might someday fade. But to think so would be naïve: studies prove (ad nauseam) that beauty counts, even when we think it shouldn't matter. An ultrasound technician at New York Presbyterian Hospital told me that once expectant parents get assurances about their unborn baby's health, they ask questions about its nose. "They want to learn that the baby is OK, and then they want to look at the nose," she said. "Sometimes people get really upset and say, 'Oh, no, he got your nose!' We're talking about a sixteen-week-old fetus." But it's not so mysterious. Mothers coo more to pretty babies than to plain ones, and we go out of our way to help the better-looking among us. Both women and men admire symmetry; they broadcast good genes—biological safety. As for the nose, well, as anyone with a less-than-perfect one can tell you, there is no such thing as it "not mattering."

If Americans, by birthright, pursue happiness, so too do they pursue beauty. And rhinoplasty—with all its pain, risks, dubious results, and expense—will prevail, as a rite of passage, an icon, and a metaphor.

Epilogue

After nearly three years of living with the nose and its intricacies, one thing, at least, has become as obvious as the nose on my—or anyone else's—face. As little understood as it is, the nose is an organ that is the core of everything imaginable in life. We undergo significant pain and cost to change its appearance. The billions we spend to cure sinus disease testify to our often-fruitless search for good health. We roll deodorant on our armpits and splash perfume on our wrists and necks in an attempt to please it. We tempt it with tantalizing aromas, from vintage merlot to sizzling prime rib. We attract mates with enigmatic chemicals detected by a tiny, mysterious organ lodged deep within it, if pheromone theory turns out to be correct.

The process of writing this book was, in some ways, a gift.

For the first time, I came to truly understand the centrality, and magic, of smells in my world. At some point, it became clear to me that I could continue my research for years, as the recorded history of the nose stretches back to the beginning of time. Even so, its starting point for me is a clear one, snapping into focus on a snowy February night in 1998.

It was my husband's birthday, and my daughters and I were baking an orange cake. I had grated the peel and squeezed out the juice from two huge navel oranges, but to me, the extra flourishes were meaningless. The essence of the fruit was lost, its sunny aroma as silent as a white sheet of paper. Why not just use Betty Crocker? I thought. I stood, sullen as a teenager, as the girls broke eggs and mixed in flour. The dining room table was set; red tulips drooped seductively from a vase, and a fire crackled in the living room. Moisture gathered on the kitchen windows, and the girls giggled as they stirred.

My annoyance mounted to outright meanness. I scolded them for getting flour on the floor—they were five and two and a half at the time—and I snatched the mixer from my older daughter as it clinked against the Pyrex bowl. (Although science has disproved the so-called "Helen Keller theory," my inability to smell seemed only to augment my other senses. Sounds, especially, seemed to grate on me in all sorts of irritating ways. Or maybe it's simply that I concentrated on them more.)

Just as I was putting the pans in the oven, my husband burst through the door and screamed at us to get out of the house. I looked at him, and set the timer. Did he want us to see the snow? I wondered dully. "The stove is leaking gas!" he shouted. "I can smell it clear out on the street! Get out of the house!"

He ushered the girls onto the deck, and shoved an anorak at me. "Get out!" he shouted. "What's the matter with you?" I stood, insanely mute, as I watched my family on the whitened lawn. Finally I went outdoors. I couldn't smell my kids, I couldn't smell myself, I couldn't even detect danger. I just stood and sobbed in the snow.

Desperate, I sought help from my allergist, an old-fashioned but up-to-date doctor who listens to his patients. He theorized that after four surgeries, the trauma to the delicate lining of my sinuses was so severe that no odor molecules could possibly make their way to the olfactory epithelium. So he wrote out a prescription for a very high dose of prednisone, 80 milligrams. I had already tried several go-rounds with lower doses, to no avail. This time, I was to taper the pills down over the course of two weeks to 5 milligrams a day. Its unpleasant side effects are legion, and the main one I experienced—irritability—added to my already sorrowful state. But by the end of the two weeks, I could smell again. This time, it wasn't just an errant whiff of garlic—it was everything, sometimes with excruciating clarity. From bad breath to fake grape bubblegum to an overheated, overchlorinated indoor pool in a mildewy 1950s YMCA, I could breathe it all in. I felt like Dorothy traveling from Kansas to the Land of Oz. Suddenly my nose had switched from "seeing" black and gray cornfields to glittering ruby slippers.

I finished this book while I was pregnant, a time during

which hormones heighten the ability to smell. In a few years, I had gone from not smelling at all to smelling too much. One particularly queasy morning I was convinced I could smell an egg cooked over easy and sprinkled with black pepper from a house nearby. The coffee on my students' breath at journalism school made me want to bolt. During a trip to Paris, a man next to me on the subway platform ate a tuna sandwich from a crisp paper wrapper as he calmly scanned *Le Monde*. Combined with the scent of overheated urine, I couldn't, and didn't, bear it. I love Paris, and I love its smells: melted Gruyère on a Croque Monsieur; chocolat chaud wafting from a café in winter; briny seafood on ice in a sidewalk display. But this time, as I struggled to keep nausea at bay, I found myself longing, ironically, for the isolation of anosmia.

Pregnancy is a misery for me, with nausea and stuffiness filling hours insomnia doesn't. When I can't sleep in normal times, I pop a Benedryl, and as I wait for it to work I come up with crazy lists intended to invite sleep—I alphabetize world capitals, or bodies of water. But in my heightened olfactory state, for a few weeks my mind fastened on smells and wouldn't let go. Had something triggered them, or was my mind—and nose—just playing tricks on me? Was it nerves? Hormones?

Stranger still, the smells that roamed into my brain each night were smells particular to a place and era: Eastern Europe of a decade prior. Perhaps not surprisingly, the odors had been accompanied by intense emotions in real time.

My husband and I were based as reporters in Poland for three years in the early 1990s, and I wasn't particularly happy there. I was especially cross about the weather (winter drags on reliably until mid-April) and the lack of light. (In December, night falls at three in the afternoon.) My husband, a native Bostonian, had no qualms about the short, dark days. In fact, he loved Central Europe, and the juxtaposition of our sentiments created flurries every now and again. But what really tested our contrasting sensibilities, I found, were smells. What drove me crazy—cheap black diesel fumes, ripe body odors on the tram in midsummer, the salty, garlicky odor of open-air butcher shops, where pigs dangled from hooks in full view—he scarcely noticed.

On restless nights a good ten years later, pregnant with my third child in my bed in New York, my nostrils couldn't shake a trip to Slovakia. I wasn't worried about myself (olfactory hallucinations can be linked to schizophrenia), but I was annoyed—the odors were rank. And like dreams, they were uninvited and nonsensical, even rude. I wasn't imagining daphne, sun-dried sheets, or the smell of my husband's hair. My mind somehow stuck on a stinking trip to Bratislava on a hot week in June.

It started out like this. Early one morning after a gulped-down breakfast, we gathered for a "press conference" in the dank medieval wine cellar of an unctuous politician who was doused with cheap aftershave. There are no dictates against drinking early in the day in Central Europe, and a "hostess" in a miniskirt thrust glasses of the sweet, golden Tokaj into everyone's hand. It is considered bad form not to at least take a sip, and the taste I took landed somewhere between Tinkerbell perfume and the red-dyed sugar mix you put in a hummingbird feeder.

Next on the agenda was lunch. The waitress recommended a "light Slovakian appetizer, specialty of Bratislava." She returned some minutes later with a heaping platter of hot, fried cheese. Globs of grease settled on the plate, and to my mind, both then and now, it smelled like a carton of Kentucky Fried Chicken someone had accidentally left in the car.

That evening, my husband suggested a romantic little stroll on the banks of the Danube. By that time, I, and my nose, had had it. "No way," I said. "Why not?" he asked. "The river stinks," I said. He laughed out loud. "What?" "The river stinks," I repeated. "It does *not,*" he said. This conversation deteriorated into one of our most irrational arguments. "That's my whole problem here!" I recall shouting. "The river stinks, and you can't smell it!"

Modern science has yet to come up with a full explanation for why millions of people like me lose their sense of smell, or why a far smaller and luckier group manage to regain it. (Imagined odors are another question still.) But my journey through smells—lost, regained, and imagined—gave me an unusual appreciation of the secret world of neurons, receptors, and molecules, as well as the scientists who try to puzzle it all out.

My voyage through science, medicine, literature, and commerce has been a humbling one. If one theme seems clear, it is how little we really know. If history is any guide, a large percentage of what we now believe to be true about the nose will turn out to be uninformed or flat-out wrong. The sequencing of the human genome and the understanding of how specific genes affect smell is only in its infancy, as is the true grasp of how odor influences our memories, thoughts, and behavior.

As I learned, I began to sense the truth behind the aphorisms in many cultures suggesting that the nose can be followed to knowledge. The French compliment a person's intelligence by saying, *"Il a un bon nez,"*—"He has a good nose." The Germans have an expression, *"naseweise,"* literally, "nosewise." And in English we have our own maxim: The nose knows. While that may well be true, unlocking its secrets is still beyond us. The nose is aware of far more than we know—yet.

Notes

Chapter 1: Memoir of a Nose

1 The Egyptians believed that it was good luck to sneeze after looking at the sun; the Romans and Greeks found it an augur of good fortune as well. In Greek mythology, Prometheus fashioned a clay image of a man, then seized fire from the sun and applied it to the statue's nostrils. It sneezed, and became a living man. Aristotle believed a hearty sneeze was an "emotion of the brain," and was evidence of the brain's vigor. "He who hears it is honored," he wrote.

Saying "Bless you" after a person sneezes appears to stem from an ancient Jewish custom. Today, we say it as a reflexive courtesy, but in ancient Israel, a sneeze was a dire matter. According to the Talmud, from the time that God made Heaven and Earth a person never became ill. There is no mention of sickness in the Bible until Jacob hears of his father Isaac's: "Behold, your father is ill" (Genesis 48:1). Jacob asked God to give man warning of his impending demise, so that he might have time to repent. And so a sneeze, then, came to symbolize a death knell, prompting those who heard the sneeze to call out blessings—"Gesundheit" (to your health) or "God bless you."

This tradition traveled with the Diaspora. By the time the Black Plague struck Europe in the fourteenth century, Pope Gregory VII decreed that healthy people who heard a sneeze—one of the first signs of affliction was a violent sneezing fit—should utter "God bless you" in the hope of staving off the disease. Failing that, the benediction at least granted the sick a chance to repent.

Chapter 2: Centuries of Stench

1 Morris, Edmund, *Scents of Time* (New York: Metropolitan Museum of Art, 1999), p. 22.

2 Lise Manniche, *Sacred Luxuries: Fragrance, Aromatherapy, and Cosmetics in Ancient Egypt* (Ithaca, NY: Cornell UP, 1999), p. 33.

3 Pilch, John J., *The Cultural Dictionary of the Bible* (Collegeville, MN: The Liturgical Press: 1999), pp. 14–20.

4 Nard, or spikenard, is an aromatic imported from the Himalayas. It is a member of the valerian family, which also includes the more familiar (and heavenly) heliotrope.

5 For more history on the ancient baths, see Mikkel Aaland's book, *Sweat*. Published by Capra Press in 1978, it is out of print, but excerpted at www.cybohemia.com /Pages/massbathing.htm.

6 According to legend, women washing clothes in the Tiber River discovered that the clay riverbed beneath Mount Sapo had special properties—this attributed, naturally, to gods pleased with animal sacrifices made on the hilltop. What is more likely is that the animal fats and alkali, leached out of the ashes by winter rains, streamed into the Tiber, where it accumulated in the soil. The story is likely just legend, but *sapo* is the origin of the word "soap" in many Indo-European languages. Other legends date the discovery of soap to the Roman invasion of England, in 43 CE; there, they learned the art of soap making from Celts, who called it *saipo*.

7 LeGuerer, Annick, *Scent: The Mysterious and Essential Powers of Smell* (New York: Turtle Bay Books, 1992), p. 41.

8 Classen, Constance; Howes, David; Synnott, Anthony, *Aroma: the Cultural History of Smell* (New York: Routledge, 1994), p. 49.

9 This passage by Plautus is noted in *Aroma,* p. 49. The book is a great portrait of smell and its meaning around the world.

10 St. Jerome, *Select Letters,* translated by F.A. Wright (Cambridge, MA: Harvard University Press, 1975), p. 183.

11 Classen, Howes, Synnott, p. 130.

12 *Smithsonian Magazine,* February 1990, Vol. X, p. 66. "How a Mysterious Disease Laid Low Europe's Masses," by Charles L. Mee Jr.

13 *The Florentine Chronicle* was written by Marchione di Coppo Stefani in the 1370s and 1380s. The original appears in *Cronaca fiorentina. Rerum Italicarum Scriptores,* Vol. 30., ed. Niccolo Rodolico. Citta di Castello: 1903–13. A modern translation is available at: http://jefferson.village.virginia.edu/osheim/marchione.html.

14 Defoe, Daniel, *A Journal of the Plague Year* (New York: Signet Classics, 1963), p. 79.

15 *Ibid.*, p. 93.

16 Classen, Howes, Synnott, p. 62.

(The nursery rhyme "Ring Around the Rosy," which originated during the plague era in Britain, notes the custom of protecting oneself with herbs and flowers with the line, "A pocket full of poseys." The ring, meanwhile, referred to the darkened red swellings that appeared on the victims; "Ashes, ashes," is a corruption of "Achoo, achoo." Sneezing fits seized the afflicted—while they still had the strength to emit them.)

17 Suskind, Patrick, *Perfume* (New York: Pocket, 1991), p. 4.

18 Corbin, Alain, *The Foul and the Fragrant: Odor and the French Social Imagination* (Cambridge: Harvard UP, 1986), p. 46.

19 *Ibid.*, p. 44.

20 Hoy, Suellen, *Chasing Dirt: The American Pursuit of Cleanliness* (New York: Oxford UP, 1995), p. 24.

21 The PBS television show "The 1900 House," which aired in the United States during the spring of 2000, gave the following statistics: in Britain in 1900, there were 2,000 scalding deaths per year; in 1999, there were 11.

22 Hoy, p. 35.

23 Hoy, p. 48.

24 Hoy, pp. 171–173.

25 Leopold Senghor, "New York." From Leopold Sedar Senghor, *Selected Poems,* translated and introduced by John Reed and Clive Wake. (New York: Atheneum, 1969).

Chapter 3: The History of Science

1 Weir, Neil, *Otolaryngology, An Illustrated History* (London: Butterworths, 1990), p. 54.

2 The Digital Lavater, published online by http://www.newcastle.edu.au/department/fad/fi/lavater/lav-intr.htm

3 Charles Darwin, *The Autobiography of Charles Darwin, 1809–1882,* ed. Nora Barlow (New York: W.W. Norton, 1993), p. 72.

4 Jabet, George, *Notes on Noses* (London: Richard Bentley, 1859), pp.2–3.

5 *Ibid,* pp. 9–11.

6 *Ibid,* pp. 121–127.

7 Morantz-Sanchez, Regina, *Conduct Unbecoming a Woman: Medicine on Trial in Turn-of-the-Century Brooklyn* (New York: Oxford UP, 1999), p. 117.

8 Mackenzie, J. R., *American Journal of Medical Science,* LXXXVII, 1884, pp. 360–365.

9 Morantz-Sanchez, Regina, *In Science and Sympathy: Women*

Physicians in American Medicine (New York, Oxford: Oxford UP, 1985), pp. 116–117.

10 Ehrenreich, Barbara, and English, Deirdre, *For Her Own Good: 150 Years of the Experts' Advice to Women* (New York: Doubleday Anchor, 1979), pp. 122–123.

11 E-mail with Dr. Steven Reisner, Freud scholar and professor of psychology at Columbia University, March 5, 2001.

12 Masson, Jeffrey Moussaieff, *The Assault on Truth: Freud's Suppression of the Seduction Theory* (New York: Pocket, 1998), p. 57.

13 *Ibid*, pp. 62–66.

14 Mayer, Emil, *Journal of the American Medical Association,* Jan. 3, 1914, LXII:1, p. 7.

15 Obendorf, C.P., *Psychoanalysis* X, 1929, pp. 228–241.

16 Holmes, Thomas H.; Goodell, Helen; Wolf, Stewart; Wolff, Harold G., *The Nose: An Experimental Study in Reactions Within the Nose in Human Subjects During Varying Life Experiences* (Springfield, IL: Charles C. Thomas, 1950), pp. 69–77.

17 *Ibid,* pp. 30–32.

18 *Ibid,* pp. 89–91.

19 *Ibid,* pp. 96–97.

20 Holmes, et al., pp. 120–122.

21 Holmes, et al., pp. 125–135.

22 In fact, informed consent was first implemented in the United States by the U.S. Army at the end of the nineteenth century. Walter Reed, a bacteriologist, was intent on trying to puzzle out the cause of frequent outbreaks of yellow fever in the Southern states and the Caribbean. But it was not until an outbreak in Havana threatened American soldiers during the Spanish-American War that scientists worked in earnest to solve it. And so, with the support of the Surgeon General, Reed headed the Yellow Fever Commission on a mission to Cuba, where an epidemic was decimating both military and civilian populations. While many theories abounded, Reed was particularly intrigued by a fifty-year-old suggestion that insects spread the disease. Cold winds seemed to halt outbreaks, and Reed wondered if the weather interrupted the breeding patterns of mosquitoes. Others, however, believed the disease was spread by the spittle or blood of infected victims.

Since no known animals contracted the virus, Reed asked for healthy volunteers to test his hypothesis. Several Spanish immigrants participated in the experiments, but the majority of volunteers came from nearby troops. Reed was authorized to pay each volunteer $100 in gold apiece, which to a poor Spanish immigrant or an underpaid army private was considerable incentive. In a consent form—possibly the first of its kind—Reed emphasized that contracting the often fatal disease was highly likely. But because his experiment would be in a controlled environment with medical care available immediately to those who fell ill, Reed suggested that the chances of recovery were far greater than those in a remote camp.

One set of "volunteers" slept in a screened chamber on the clothes and sheets soiled with the blood of yellow fever patients. Another was sequestered from the patients' fouled linens and clothing but exposed to the mosquitoes that had been allowed to bite people ill with yellow fever. None of those who slept on the contaminated items became sick; those bitten by the mosquitoes contracted the virus immediately.

Curiously, such consideration was not extended to *sick* patients until after World War II. As the Nuremberg trial got underway, the American Medical Association enlisted Dr. Andrew Ivy, a leading medical researcher, to draft a hasty policy for medical experimentation. Ivy's proposal included the following three requirements: voluntary consent of the person on whom the experiment is to be performed must be obtained; danger of each experiment must be previously investigated by animal experimentation; and the experiment must be performed under proper medical protection and management. At a December 1946 meeting, the AMA voted its approval of Ivy's draft.

According to a report published in 1995 by the Advisory Committee on Human Radiation Experiments (http://tis.eh.doe.gov/ohre/roadmap/achre/chap2_2.html), some 70 percent of American physicians belonged to the AMA. Association members received a regular subscription to the *Journal of the American Medical Association*, but in fact, these rules were not published prominently. They were set in small type along with a variety of other miscellaneous business items in the lengthy published minutes of the meeting. Only an exceptionally diligent member, or one with a special interest in medical ethics, is likely to have located this item.

23 Dr. David Kennedy, the chief of otolaryngology at the University of Pennsylvania, cited Wolff's work on stress and sinus disease at a Washington, D.C., conference called "The Nose 2000," in an address on September 22, 2000.

Chapter 4: Nagasaki Up the Nose

1 Figures on childhood deafness prior to the late 1950s are impossible to come by, according to the National Institutes of Health and the Centers for Disease Control and Prevention. The 4 million figure comes from the *Saturday Evening Post* article; whether it is reliable or not is debatable. The article was published on August 14, 1948.

2 The Johns Hopkins University School of Medicine and School of Hygiene and Public Health, to Federal Security Agency, Public Health Service, National Institutes of Health, July 1948 ("The Efficiency of Nasopharyngeal Irradiation in the Prevention of Deafness in Children,

Notice of Research Project, Grant No. B-19," ACHRE No. HHS No. 092694-A).

3 These comments were first reported by Molly Rath in a Baltimore *City Paper* feature called "Fair Treatment," which ran on November 4, 1997. It is available online at: http://members.aol.com/radproject/city_paper.html.

4 Proctor, Donald, *The Tonsils and Adenoids in Childhood* (Springfield, IL.: Charles C. Thomas, 1960), p. 18.

5 http://tis.eh.doe.gov/ohre/roadmap/achre/chap18_1.html.

6 This is from the March, 29, 1996, edition of the CDC's weekly morbidity and mortality report: http://www.cdc.gov/mmwr/preview/mmwrhtml/00040719.htm.

7 The musician Frank Zappa, who was born in Baltimore, was also irradiated as a child. He describes the ordeal in his autobiography, *The Real Frank Zappa Book.* He recalled a doctor stuffing "a long wire thing" up his nose. Zappa wondered later if his "handkerchief is glowing in the dark." (New York: Poseidon Press, 1990, p. 20.)

Chapter 5: What Smells?

1 Brodal, A., *Neurological Anatomy,* 3rd. ed. (New York: Oxford, 1981), p. 640.

2 Interview, Dr. Barry Davis, director of the Taste and Smell Programs at the National Institutes on Deafness and Communication Disorders, which oversees research and funding for olfaction at NIH, February 5, 2001.

3 Chemosensory scientists study olfaction, of course, but also gustatory, or taste cells, which react to food and drinks. We can commonly identify four basic tastes: sweet, sour, bitter, and salty. In the mouth these tastes combine with texture, temperature, sensations, and odor to produce flavor. These cells, which line the surface of the tongue, send taste information to their nerve fibers. The taste cells are clustered in the tiny bumps on the tongue—taste buds. Another mechanism is called the "common chemical sense," which contributes to smell and taste. In this system, thousands of nerve endings—especially on the moist surfaces of the eyes, nose, throat, and mouth—are responsible for the sensations we get when we feel the sting of rubbing alcohol, the burn of Scotch, the cool of peppermint, or the irritation of hot peppers.

4 Classen, Howes, Synnott, pp. 98–100.

5 Sacks, Oliver, *The Man Who Mistook His Wife for a Hat and Other Clinical Tales* (New York: Harper Perennial, 1990), pp. 156–158.

6 Richard L. Doty, the director of the University of Pennsylvania's Smell and Taste Center, developed a multiple-choice smell test—an eye exam for the nose—in the early 1980s. The test, which is called the University of Pennsylvania Smell Identification Test, is used to determine a number of

disorders, including degenerative neurological diseases such as Alzheimer's and Parkinson's.

7 Rodriguez, Ivan; Greer, Charles A.; Mok, Mai Y.; Mombaerts, Peter, *Nature Genetics,* September 2000, Vol. 26, No.1, pp. 18–19.

8 Jacob, Suma, and McClintock, Martha, *Hormones and Behavior,* Vol. 37, No. 1 (Feb. 2000): 57–78.

Chapter 6: The Recreational Nose

1 "Mental Sequel of the Harrison Law," *New York Medical Journal,* 102 (May 15, 1915): 1014.

Chapter 8: The Sinus Business

1 At the Washington conference, a faculty disclosure sheet listed the companies for which various speakers served as consultants. David Kennedy was listed as a consultant for three pharmaceutical companies, as well as for two companies, which use the global positioning systems in operating rooms. He receives royalties for one of them as well. Sherris is a consultant for Bayer; Ponikau advises for Schering; Kern does not consult.

2 While stress is clearly a factor in illness, in sinusitis, unlike heart disease, it is all but impossible to quantify. Only anecdotal evidence links the two. And as for pollution, Ponikau and others point to the landmark study done in the early 1990s by Dr. Erika von Mutius at the University Children's Hospital in Munich. Von Mutius compared asthma and allergic rhinitis rates of 5,600 children in polluted Dresden to that of state-of-the-art clean Munich, and found that the asthma rate among West German children was significantly higher than it was in East Germany. Von Mutius was so surprised at the results that she checked and double-checked all the data entries. She was convinced they had made a mistake, but again and again the numbers were the same.

Indeed, between 1975 and 1995, asthma rates in children doubled in many parts of the world, and the rates for adults also rose sharply. But pollution does not appear to be the culprit: when people from the Polynesian island of Tokelau move to New Zealand, their chances of getting asthma double. Similar increases have been seen among Filipinos moving to the United States, and Asians from East Africa moving to Britain. Chinese people in Taiwan, who have stayed in the same place, but gradually adopted a Westernized lifestyle, now have eight times more cases of childhood asthma than they had in 1974. In Ghana, wealthy urbanites have more asthma than do the poor in the same cities; in remote villages, there is little or none of the disease. Asthma researchers say that "Westernized life" is to blame, but that does not account for high rates of asthma in Brazil and Peru, where it strikes all socioeconomic groups. In his book called *Asthma: The Complete Guide,* Dr. Jonathan Brostoff, a professor of Allergy and Environmental

Health at University College London Medical School, says that air pollution is usually blamed, but the case against it "simply doesn't hold up." Although it can make asthma worse for people who already have the disease, and may produce a *small* increase in the number of people developing asthma, there is no way that air pollution is the major cause of the asthma epidemic. Some places with extremely clean air, such as New Zealand, have very high rates of asthma. (Healing Arts Press, Rochester, VT, 2000, pp. 78–85.)

3 Mayo has received patents on the two antifungal medications it prescribes to patients.

Chapter 9: Smells Like Money

1 Hoy, Suellen, *Chasing Dirt: The American Pursuit of Cleanliness* (New York: Oxford UP, 1995), p. 146.

2 Friedan, Betty, *The Feminine Mystique* (New York: Bantam Doubleday, 1984), pp. 209–219.

3 Green, Annette, and Dyett, Linda, *Secrets of Aromatic Jewelry* (Paris: Flammarion, 1998), pp. 137–138.

4 Many of the smell researchers interviewed for this book have received fund grants.

Chapter 10: The War on Stink

1 Mark Magnier, "The Smell of Aging," Los Angeles *Times,* p. 1, July 14, 1999.

2 Classen, Constance; Howes, David; Synnott, Anthony, *Aroma: The Cultural History of Smell* (New York and London: Routledge, 1994), pp. 182–183.

3 Classen, Howes, Synnot, p. 184.

4 From a pamphlet distributed to foreign students at the Rochester Institute of Technology, Rochester, NY, 2002.

Chapter 11: From the Bronx to America

1 Gilman, Sander, *Making the Body Beautiful: A Cultural History of Aesthetic Surgery* (Princeton: Princeton UP, 1999), p. 68.

2 Holden, Harold M., *Noses* (New York: World Publishing Company, 1950), pp. 203–204.

3 Gilman, p. 57.

4 Simons, Robert L., *Coming of Age: A Twenty-fifth Anniversary History of the American Academy of Facial Plastic and Reconstructive Surgery* (New York: Thieme Medical Publishing, 1989), pp. 2–3. It is unclear, Simons says, whether the patient described here was indeed Jewish, but scholars conclude from Jacob's wording that he was.

5 *Ibid,* pp. 3–6.

6 Holden, Harold M., *Noses* (Cleveland and New York: World Publishing Company, 1950), pp. 106–108.

7 Interview, Robert L. Simons, June 19, 2001.

8 Ofodile, Ferdinand, *Annals of Plastic Surgery,* Vol. 32, I (Jan. 1994): 25.

9 These figures were provided by the American Academy of Plastic Surgery in Arlington Heights, IL. Statistics from previous years are not appropriate comparisons, because 2000 was the first year the organization tracked figures from all doctors who perform the procedure. It can be performed by an ENT, a general plastic surgeon, or a facial plastic surgeon. In 1992, for example, the academy listed 50,175 rhinoplasties, as reported by facial plastic surgeons. But officials in Arlington say that the overall figure has remained fairly steady throughout the decade.

Epilogue

1 During World War I, American cotton companies endeavored to create a gas mask that would protect soldiers from poison gas attacks. Eventually, they did, using a supply of crepe-like cotton as a buffer allowing the soldiers to breathe. Once the war ended, the companies had a surplus of the soft cotton that they pondered how to use.

At first, product designers used the cotton for the first commercial sanitary pads. Wartime nurses were the first to use the cotton for menstrual blood; they discovered that it was much easier than the traditional flannel rags, which had to be washed and reused. Prim mores of the era were a huge impediment to selling the pads, however: magazines refused to advertise them, and stores balked at carrying them.

By 1919, an answer for the surplus arrived in the name of the automobile. Americans owned 6 million cars, which were mostly open-air "touring automobiles." While men wore goggles and mustaches, women got windburn on their noses and cheeks, and began to use cold cream nightly as a remedy for their red skins. In the morning, however, they needed something to remove the oily residue. Towels were an obvious choice, but since few homes had modern washing machines, the idea of disposable tissues made perfect sense. Officials at Kimberly Clark, the cotton giant, instantly saw a market. Soon, American drugstores were filled with ads—and boxes—of Kleenex tissues. The name, Kleenex, derives from its intended use, with an "ex" added to cement brand loyalty.

Meanwhile, Ernst Mahler, a Kimberly Clark executive who suffered from hay fever, began filching his wife's tissues for his own runny nose. The handiness of the tissues, combined with lingering fears of the flu epidemic of 1918–19, helped spawn an instant product: nasal tissues. By the late 1920s, the company began to market its "cleansing" product as a hygienic one. Ads for Kleenex showed a sneezing child spewing mucus into a tissue:

"Colds fill handkerchiefs with germs—boiling water fails to kill them! A handkerchief used once during a cold is unfit to be used again. Avoid Reinfection! Use Kleenex disposable tissues."

Today, Americans go through 200 billion Kleenex a year. Puffs, Kleenex's top competitor, in 1990 introduced an improved version of disposable tissues, laced with lotion and aloe for sore noses. Together, they account for some $1.2 billion in revenues.

Acknowledgments

This book, no ordinary project, took advantage of many talented souls. Thanks are due the following physicians for their patience with a layman: Joseph Gogos, Richard Axel, Richard Garvey, Ferdinand Ofodile, and Robert Henkin; at the Mayo Clinic, Jens Ponikau, David Sherris, Eugene Kern, Thomas Gaffey, and Thomas McDonald, as well as their Austrian colleagues, Heinz Stammberger, Walter Buzina, and Hannes Braun.

Thanks are also due Rachel Sarah Herz, Stewart Farber, Tyler Lorig, Luis Monti-Bloch, David Berliner, Richard Doty, Annette Green, Theresa Molnar, Mike Norwood, Adolfo Coniglio, and Rabbi Lawrence Perlman. My own doctors, Clark Huang, James Pollowitz, Michael Tom, and Scott McNamara, have been exceedingly tolerant of a know-it-all patient. Thanks to Miriam Reinharth, Diane Solway, Rita Beamish, Michelle

Jackson, Jon Jackson, Mitchell Glaser, Wendi Glaser, Nina Martin, Jane Gross, and Edward Engelberg for their highly capable eyes.

My agents, Glen Hartley and Lynn Chu, played an important role in creating this book, from the initial idea to the galleys. They are deeply loyal and decent human beings. Rosemary Ahern is an old-fashioned book editor who actually reads and edits manuscripts. Thanks to her lyricism, this book read much better after she got done with it. My husband, Stephen Engelberg, read—and improved—this book in many drafts, even when he was on deadline with his own. He is, in all senses of the word, a mensch—and a brilliant one at that. And while book writing gives you time with your family, it also often takes away: from bedtime, from soccer games, from swimming dates (made and broken, made and broken) on summer afternoons. Thanks to my older daughters, Ilana and Moriah, for understanding.

Mostly, though, I'd like to thank my parents, Steve and Virginia Glaser, and my grandmother, the late Violet Claypool, for creating such a lively world of smells. It travels well.